Changing the Way America Farms

VOLUME 12 IN THE SERIES

OUR SUSTAINABLE FUTURE

Series Editors

Cornelia Flora
Iowa State University

Charles A. Francis
University of Nebraska–Lincoln

Paul Olson
University of Nebraska–Lincoln

Neva Hassanein

Changing the Way
America Farms

Knowledge and Community
in the Sustainable
Agriculture Movement

University of Nebraska Press, Lincoln and London

Library of Congress
Cataloging-in-Publication Data. Hassanein, Neva,
1963–
Changing the way America farms:
knowledge and community in the sustainable
agriculture movement / Neva Hassanein.
p.   cm. –
(Our sustainable future) Includes bibliographical
references and index.
ISBN 0-8032-7321-5 (pa. : alk. paper)
1. Alternative agriculture – United States.
2. Sustainable agriculture – United States.
I. Title.   II. Series.
S494.5.A65H37   1999
338.1′62′0973 – dc21
99-18477 CIP

To the innovative farmers who have shared
their knowledge and communities with me,

and in memory of Mike Cannell

Contents

Illustrations

Acknowledgments

This book would not have been possible without the generosity of the members of the Ocooch Grazers Network and the Wisconsin Women's Sustainable Farming Network, the two groups that are the focus of this study conducted during the early 1990s. I am deeply grateful to the members of these networks for welcoming me to their events and for teaching me so much in so many ways. Individuals interviewed by me and quoted extensively in these pages are identified by their real name or by a pseudonym, depending on their own preference.

This work is dedicated to these innovative farmers and to the memory of one of them, Mike Cannell. At the age of fifty-five, Mike was killed in a machinery accident while helping a neighbor unload corn in December 1996. Described as "one of the brightest lights in rural Wisconsin," Mike was a strong and generous leader in the sustainable agriculture movement, and, as the reader will discover in the coming pages, he spoke with a rare eloquence about the need to build healthy and socially just rural communities. On a personal level, Mike enlightened and challenged me. I wish he were here to give me his characteristically frank opinions on this work; no doubt, he would have improved it.

My graduate school adviser and friend, Jack R. Kloppenburg Jr., was also a key ingredient in making this book a reality. During my years at the University of Wisconsin – Madison when this study was conducted, Jack repeatedly encouraged me to write a book, a thought that simultaneously intrigued and terrified me. He underscored his confidence in me by providing me with the intellectual tools I needed and by making available generous financial support through grants from the College of Agricultural and Life Sciences at the University of Wisconsin and from The Pew Scholars Program in Conservation and Environment. I am truly thankful to Jack and to these institutions.

Others at the University of Wisconsin supported my work. I very much appreciate the advice and feedback of Claudia Card, Jane Collins, Jess Gilbert, and Virginia Sapiro. I would also like to express my sincere appreciation for the institutional support of the Institute for Environmental Studies and the Department of Rural Sociology.

I appreciate the interest and assistance of the University of Nebraska Press. For their thoughtful and useful reviews of the manuscript, I am indebted to Patrick H. Mooney, Charles A. Francis, and the two anonymous reviewers whom Charles recruited on very short notice.

Reaching this milestone puts me in the happy position of being able to reflect on the experiences that have shaped my perspective and the personal supports that enabled me to complete this book. Three of my undergraduate teachers – Micere Mugo, Leslie King, and Glenn Harris – showed by example how one can be both a scholar and an activist. The farmers, ranchers, and community organizers I worked with during my years at the Northern Plains Resource Council in Montana greatly deepened my understanding of contemporary rural issues and social movements. More recently, my colleagues at the Northwest Coalition for Alternatives to Pesticides have supported my occasional need to abandon my duties there during the final stages of completing this work.

My parents, each in their own way, have helped make it possible for me to move along this path in life, and I am thankful for their support. Friends Norma Grier, Janice John, Lori Maddox, Ellen Rifkin, and Njanja Rwenji provided much-needed encouragement. And I fondly remember discussions beside Lake Mendota with Marcy Ostrom and Eva Jensen, who helped me to navigate graduate school and to shape the research questions I pursued.

Finally, my deepest gratitude goes to Lita Furby, who has sustained both me and this project in countless ways from beginning to end. Thank you.

Changing the Way America Farms

ONE

A Social Movement
to Change Agriculture

"I basically want to change the way America farms," said Faye Jones when I interviewed her about her participation in one of the sustainable farmer networks that constitute the focus of this study. With that simple sentence, Faye captured the central goal of a social movement that seeks to establish alternatives to the conventional food system in the United States and which has steadily gained momentum in recent years. The movement includes people who hold different, but associated, visions of what a transformed agriculture might include – whether it be described as ecological, biodynamic, regenerative, low-input, organic, permanent, natural, community-supported, alternative, or sustainable. Movement participants work to realize these visions not only in farmer networks such as the one Faye belongs to in Wisconsin but also in such places as natural food stores, community gardens, advocacy organizations, farmers' markets, and research centers throughout the country. Whatever the differences are between proponents of these numerous schools of thought and activism, collectively these efforts have been recognized as the "sustainable agriculture movement."

A defining characteristic of social movements is collective activity that is intended to promote or resist social change and occurs with some degree of organization and continuity outside of established institutional channels (Benford 1992). The recently formed Campaign for Sustainable Agriculture attests to the existence of an unprecedented level of collective activity oriented toward transforming agriculture throughout the United States. The campaign includes nearly six hundred organizations representing family farmers, consumers, environmentalists, fish and wildlife interests, animal protection supporters, the religious community, farmworkers, community food activists, and others (Lawrence 1995). As this diversity indicates, the sustainable agriculture movement is not a thing in and of itself. Like other social movements, the effort to establish a more sustainable agriculture is a dynamic, multidimensional process involving various people situated in particular places, who create and implement assorted strategies, participate in diverse forms of action, and encounter a variety of obstacles and opportunities.

Underlying the diverse strands of thought and activism within sustainable agriculture is the message that so-called conventional agriculture does not represent the only or the best possible model for producing, processing, and distributing food and fiber. There are alternatives. To realize these alternatives, however, advocates of a sustainable agriculture have found that one element is critical: knowledge. In other words, achieving an environmentally sound, economically viable, socially just system of food production and distribution requires knowledge about agricultural ecology, economics, and social relations. Often as a matter of necessity, then, many advocates of sustainable agriculture have devoted considerable effort to addressing the "knowledge question." Their efforts have included a variety of strategies geared toward improving knowledge about sustainable agricultural practices and concepts.

Changing the Way America Farms explores one such strategy. Because of the paucity of information about practices that deviate from the dominant agricultural system, farmers and rural advocates from Arkansas to Maine to the state of Washington have organized a growing number of locally based farmer networks during recent years. These networks facilitate the creation and exchange of knowledge useful to alternative farmers. The term *knowledge* refers to both substantive and technical information about specific topics and the ideological assumptions that shape how such substantive information is constructed and exchanged. Knowledge is not usually the first thing that comes to mind when one thinks of politics and social movements, but Hilary Wainwright (1994) and Ron Eyerman and Andrew Jamison (1991) have encouraged scholars and activists to recognize that social movements often include processes through which challenges to dominant knowledge claims are expressed and competing knowledge claims are generated and diffused. Without knowledge it can be difficult, if not impossible, to take effective action toward achieving a movement's goals. This book examines the way knowledge is created and exchanged in farmer networks in order to shed light on the role knowledge-related activities play in achieving the social change goals of the sustainable agriculture movement.

Goals of the Movement for a Sustainable Alternative

In the United States, the term *sustainable agriculture* emerged in the early 1980s as a potent symbol of the movement, and it has been a central concept stimulating critical discussion regarding agriculture during the last twenty years (Youngberg, Schaller, and Merrigan 1993). A wide variety of interests and concerns have been expressed in the context of sustainable agriculture. Although the boundaries of any social movement are diffuse across time and space, participants share a sense of being on the same side of a social conflict and an orientation toward some very general goals for social change (Diani 1992). As a result, the most useful definitions of sustainable agriculture tend to be broad and inclusive. The one I adopt here from Patricia Allen and her colleagues (1991:37) is no exception: "A sustainable agriculture is one that equitably balances concerns of environmental soundness, economic viability, and social justice among all sectors of society."

While activists and academics have expressed a wide range of interests and concerns in the context of sustainable agriculture, precisely what constitutes "sustainability" is contested both among movement participants and between the movement and its opponents. This diversity of opinion should not be surprising because actors in social movements articulate ideas that challenge not only established arrangements but also ideas of others in the movement. According to Alberto Melucci (1985), this ongoing process of interaction and integration of aims, beliefs, and decisions is often at the core of a movement's collective identity.

If there is a common conviction among those who identify with one or more of the goals associated with the sustainable agriculture movement, it is opposition to the industrialization, corporate domination, and globalization of agriculture. More specifically, their concern is with the consequences of industrialized agriculture's focus on continually expanding production, adopting technological innovations rapidly, using hired labor efficiently, maximizing profit, and competing in a global market (Strange 1988). Proponents of sustainable agriculture consider it undesirable that a high percentage of the total agricultural production in the United States now comes from the larger, highly specialized farms on which acres upon acres of monoculture crops cover the landscape (NRC 1989). According to the 1992 Census of Agriculture, 17 percent of U.S. farms accounted for 83 percent of the sales of agricultural products (U.S. Bureau of the Census 1995). These large-scale, specialized farms typically rely on people who work the land for wages, others who manage the operations, and still others who invest in farms for profit. Such operations require large amounts of capital to purchase a wide array of inputs such as machines, fuel, pesticides, and fertilizers and to pay for the substantial amounts of labor employed. In this market-oriented food system, farms are an ever smaller link in a complicated global agribusiness chain. From 1910 to 1990 the farm share of economic activity in agriculture declined from 41 percent to only 9 percent, while the agricultural input sector and the marketing sector grew correspondingly (Bird, Bultena, and Gardner 1995). Moreover, multinational agribusiness corporations produce agricultural products, enter into production contracts with growers, process food, supply farm inputs, and market farm output (Bonanno et al. 1994).

While acknowledging that conventional agriculture offers certain benefits such as high levels of productivity, sustainable agriculturalists point to evidence that the bountiful harvest has had adverse ecological, social, and economic consequences that are destructive of both people and the land (Bird, Bultena, and Gardner 1995). The contemporary movement is skeptical about the long-term ability of the existing system to sustain itself and those who depend on it and therefore opposes industrialized agriculture. This view results from the confluence of three distinct streams of social thought and activism: agrarian, environmental, and social justice.

First, echoing this country's early agrarian ideology, sustainable agriculturalists such as Wendell Berry (1977) are often concerned with the structure of the agricultural sector as a whole and with the integration of agriculture into the wider industrial economy. Reflecting the enduring social struggle between agrarian and industrial interests that was most forcefully waged during the height of the populist movements in the late nineteenth century, many sustainable agriculturalists have opposed further reductions in the number of small farms – especially family farms, which by definition rely principally on the labor of household members – and the resulting decline in rural communities (Beus and Dunlap 1990). Proponents of sustainable agriculture typically emphasize the need for greater decentralization, for increased farm and regional self-reliance, and for reduced economic and technological dependence on the larger industrial economy. Harriet Friedmann (1993) and other advocates of community food security emphasize the need to reconnect consumers and producers and to forge new relationships around more locally responsive food systems rather than around commodity markets. This modern form of agrarianism seeks to heighten nonfarmers' awareness and appreciation of farming and to revive many traditional agrarian values – most notably, the importance placed on community – that have largely been displaced by the values of a modern industrial economy.

Second, sustainable agriculturalists also often express deep regard for the ecological integrity of the land and human beings. As Carl Esbjornson (1992) explained, the cultural roots of sustainable agriculture lie in historical and contemporary conservation, organic farming, and environmental movements. Among the most frequently expressed ecological concerns about conventional agriculture are the high rates of soil erosion and degradation; dependence on finite fossil fuel resources; pesticide contamination of land, air, water, and wildlife; risks to human health from pesticides and nitrates; decline in genetic diversity and increasing uniformity of genetic resources; inhumane treatment of animals; and loss of wetlands, native prairies, and wildlife habitat.

To address these ecological concerns, most sustainable agriculturalists favor significantly reducing or eliminating the use of synthetic farm chemicals. For example, rather than using pesticides to minimize crop loss caused by insect pests, weeds, and diseases, alternatives include such practices as biological pest control, resistant crop varieties, crop rotations, and introduction of beneficial insects. Instead of using synthetic nutrients that are immediately available to plants and highly mobile in the soil, advocates suggest that fertilization can be accomplished and soil quality improved through composting organic matter, using animal manures carefully, adding minerals, and growing and incorporating into the soil leguminous cover crops and green manures. Many sustainable agriculturalists advocate moving away from the specialized production of one or two crops and instead creating more diverse, integrated polycultural systems. Reducing use of fossil fuel energy through greater reliance on renewable resources is also a priority. In addition, as Denny Caneff (1993) argued, sustainable agriculture involves an important shift toward alternative systems of animal husbandry, such as practicing homeopathy rather than routinely using antibiotics and growth hormones, or using intensively managed grazing systems rather than feeding livestock kept in confinement. These and other practices are aimed at conserving finite resources, enhancing and using natural processes rather than suppressing them, and producing more healthy foodstuffs.

A third and relatively new dimension of the sustainable agriculture movement concerns social justice issues. Feminists, farmworker unionists, and antihunger advocates assert that food and agricultural systems need to be improved to meet the needs of all people regardless of class, gender, or race and ethnicity. Social justice concerns have not yet been fully assimilated into the movement, which has tended to focus on environmental and economic issues. Patricia Allen and Carolyn Sachs (1993:140) have contested this focus and argued that the sustainable agriculture movement must "challenge, rather than reproduce, the conditions that led to non-sustainable agriculture in the first place." Principal among these conditions is an unequal distribution of power characterized by such problems as pervasive hunger (Clancy 1993), poisoning of farmworkers by pesticides (Moses 1993), and gender discrimination in agricultural production (Sachs 1996). Addressing these and other social justice issues fuses concerns raised by advocates of environmental justice and by ecological feminists who share the central insight that environmental problems are inseparable from the social relations of race, class, and gender. Recognizing the relationship between environmental and social justice issues constitutes a challenge to the sustainable agriculture movement aptly described by Laura DeLind (1994:147): to reunite agriculture with its social context and "to address the inequities, the exploitative relationships, and the dependencies that conventional agriculture has benefitted from but has ignored."

The Knowledge Dimension in Sustainable Agriculture

Activists and academics who identify with some or all of the goals of the sustainable agriculture movement have argued that scientific research and technology have too often resulted in ecological, social, and economic problems in contemporary agriculture. Thus a pivotal question is, How can knowledge be developed and disseminated so as to facilitate progress toward a more sustainable agriculture? For many in the movement, the answer has been to try to change the public institutions of agricultural research and extension toward lines of inquiry and research methods more favorable to sustainable agriculture (Lockeretz and Anderson 1993). Such initiatives have met with some success.

Efforts to reorient agricultural science are complemented by the creation of an alternative knowledge system that functions largely, but not exclusively, outside of the formal institutions of agricultural research and at the local level. Farmers in sustainable agriculture seem to have awakened their own problem-solving, creative capacities, steadily gaining confidence in their observations and innovations. As Wendell Berry (1984) and Jack Kloppenburg (1991) have noted, many practitioners of alternative farming techniques produce their own "local" or "personal" knowledge, and such experience-based knowledge is a powerful and promising resource for creating an alternative agriculture. Yet most previous treatments of experiential knowledge in sustainable agriculture have focused on the individual farmer and neglected the host of local and regional farmer networks around the country which create opportunities for the exchange of experiential knowledge among participants. Because sustainable farming networks have been established with the specific objective of exchanging knowledge about alternative practices and ideas among farmers, it seems important to examine the possibility that the production and dissemination of knowledge is becoming a practical foundation of this social movement.

Remarkably, social movement theorists in general, and analysts of the sustainable agriculture movement in particular, have only recently begun to recognize the role that the creation and exchange of knowledge can play in efforts to bring about social change (Hassanein 1997a). Although analysts such as Fred Buttel (1993b) have addressed the knowledge question in agriculture and have identified the struggles to realize sustainable agriculture as a social movement, they have not employed social movement theory to any significant degree. Conversely, those scholars such as Patrick H. Mooney and Theo Majka (1995) who have applied social movement theory to agriculture have not explored knowledge-related issues. Perhaps this oversight is not surprising. Only recently have social movement theorists treated knowledge as a critical dimension of their analyses. In particular, Eyerman and Jamison (1991) and Wainwright (1994) have argued that experiential knowledge can be a basis for collective action and that many "new social movements" have functioned as social laboratories in which people produce and share knowledge.

A major objective of this book is to use this new development in social movement theory to extend the current understanding of experiential knowledge in sustainable agriculture. Specifically, I examined the production and exchange of knowledge in sustainable farming networks in Wisconsin, a state where such networking is particularly well developed. To accomplish this task, I used participant observation and in-depth interviewing techniques in two networks over a two-year period. One of the two groups, the Ocooch Grazers Network, consisted of dairy farmers who practice intensive rotational grazing, a technique that represents a major departure from conventional, confinement-based dairying. The second group, the Wisconsin Women's Sustainable Farming Network, was formed by and for women farmers engaged in a variety of sustainable farming enterprises. The selection of these two particular research sites permitted an extension of theoretical treatments of local knowledge that have emphasized physical place (Kloppenburg 1991) and those that have emphasized social location (Feldman and Welsh 1995) by examining these issues in actual settings.

A second objective of this research was to consider the broader implications of local-level, collective activity centered around the creation and exchange of knowledge. On the one hand, the people I have met, the stories they have told me, the groups I have participated in, and the interactions I have observed are all unique, occurring at particular times and places. On the other hand, these local farmer networks are connected to a wider effort to transform agriculture, and that connection is most clearly evident in the nearly ubiquitous need for new knowledge about agricultural practices that are environmentally, economically, and socially sustainable. As a result, local efforts can be instructive and, in many ways, inspiring for those of us committed to building a more sustainable agriculture.

Chapter 2 begins with a discussion of the principal knowledge issues that have emerged in the context of sustainable agriculture and the two major strategies pursued to address those questions. That chapter also lays out the theoretical framework and the specific research objectives of this project. Next, in chapter 3, I describe the organizational landscape of the sustainable agriculture movement in Wisconsin, introduce the two networks studied, and detail my methodological approach. Chapter 4 presents a case study of the Ocooch Grazers, and chapter 5 focuses on the Women's Network. For each of these networks, I describe and analyze how the group functioned as an informal social movement community, how network members generated "local" and "personal" knowledge, how that experiential knowledge became socialized as network members exchanged practical information and ideas, how interpretive frameworks undergirded the knowledge exchange process, and how the networks functioned as supportive communities. Chapter 6, compares and contrasts the two networks and considers the lessons that these case studies suggest for the sustainable agriculture movement.

TWO

Knowledge Questions in the Sustainable Agriculture Movement

In 1950 the United States Congress held hearings on chemicals in food at which J. I. Rodale was the sole proponent of organic agriculture called to testify. He received a thorough trouncing as scientists refused to acknowledge his claim that chemical fertilizer could have harmful effects. According to Suzanne Peters (1979:258), Rodale's testimony was turned "into a quasi-prosecution, a sarcastic cross-examination of his credentials and professional credibility." His lack of scientific training was a substantial issue at the hearing, and his opinions were summarily dismissed in the congressional committee's report. Unable to make headway with agricultural scientists, Rodale renewed his own efforts into researching and popularizing alternatives to chemically based agriculture, most notably organic gardening and farming.

Nearly forty years later the premier research arm of the federal government, the National Research Council (NRC) published *Alternative Agriculture* (1989), a report that cataloged the now well-documented environmental problems of conventional agriculture and noted the dearth of scientific research into practices that might minimize or avoid those problems. Moreover, the report praised the innovative farmers who developed viable alternative farming practices during the many years of scientific and corporate neglect of such techniques. The NRC (1989:23) concluded that "today's alternative farming practices could become tomorrow's conventional practices, with significant benefits for farmers, the economy, and the environment."

The juxtaposition of congressional dismissal of Rodale's opinions in 1950 and the NRC's "scientific" endorsement of alternative farmers and their practices forty years later illustrates that knowledge issues have been central to the struggles of the sustainable agriculture movement. In particular, activists and academic critics have argued that the formal institutions of scientific research and the technologies they generate have been major contributors to many of the economic, ecological, and social problems in contemporary agriculture. Accordingly, individuals and organizations have worked diligently to reorient agricultural science toward lines of inquiry and research methods more appropriate to the complexities of sustainable agriculture. Indeed, the NRC report is evidence that sustainable agriculturalists have made progress after what Kloppenburg (1991:522) called "a long sojourn in the wilderness of scientific marginality."

In addition to trying to reform agricultural research policy, another major project of the sustainable agriculture movement has been to develop and disseminate knowledge in the form of new ideas, practices, and organizational principles. Farmers and other advocates of sustainable agriculture have created an alternative knowledge system that functions primarily outside of the dominant institutions of agricultural research and extension. Indeed, it was to farmer-generated knowledge that the NRC had to turn to learn about alternative practices because conventional science had little to offer. The NRC committee dedicated nearly half of its lengthy report to eleven case studies of innovative farmers and the wide-ranging alternative farming systems they had developed. Questioning the roles that science, technology, and farmer-generated knowledge play in agriculture has been one of the most prominent and enduring themes in the contemporary sustainable agriculture movement and its historical predecessors.

In this chapter, I discuss first how the formal institutions of agricultural science and technology transfer gained a perceived superiority over farmers' experientially based knowledge. Next, I explore the principal knowledge questions raised in the sustainable agriculture movement and the main strategies that participants in the movement have pursued to address those questions. To articulate a framework for interpreting the production of knowledge as a form of social action and to set out the major research questions of this study, I draw on recent contributions to social movement theory as well as other key insights regarding the power of science and knowledge in modern society. In these ways, I present both the broader context and the concepts that guide my analysis of Wisconsin farmer networks in subsequent chapters.

Farmers' Knowledge and the Rise of Agricultural Science

For nearly all of agricultural history, farmers and craftspeople have produced the knowledge necessary to farm. Farmers have tilled the soil, and they have domesticated plants and animals. Farmers have selectively bred livestock to perform a variety of services and provide a range of products.

From each year's harvest, farmers have selectively saved seeds and, in this way, created more productive, genetically diverse, and locally adapted cultivars. Farmers and craftspeople have invented and fashioned various mechanical tools, and they have built elaborate irrigation systems. In the process of producing crops, farmers have produced knowledge as well. This farmer-generated knowledge shaped the practice of farming through the vast majority of the world's agricultural history.

When agriculture became a commercial enterprise and its development accelerated rapidly in the United States during the early nineteenth century, farmers continued to be the primary source of agricultural knowledge. That knowledge was poorly developed, however, because during this expansionist period farmers tended to think that prime agricultural land was abundant and that resources were infinite. Rather than learn to maintain soil fertility and productivity through improved cultural practices, commercial farmers and plantation owners typically focused on rapidly increasing production; they found it more profitable in the short run to use land until it was depleted and then to abandon it (Busch and Lacy 1983). Some groups were important exceptions; for example, the environmental historian Joseph Petulla (1977) pointed out that Pennsylvania German farmers had long used soil-building techniques such as incorporating animal manures and red clover, making applications of lime and gypsum, and rotating crops. But for the most part during the westward migration, settlers, as Esbjornson (1992: 23) put it, "had no time for developing the long-term knowledge and wisdom necessary to evolving a loving symbiosis with their new environment."

It became clear during the nineteenth century that the improvement of agriculture depended on the improvement of agricultural knowledge. According to Douglas Hemken (1995), however, debates raged between three broad visions of how and by whom the necessary knowledge should be developed and disseminated. One view held that individual farmers could produce knowledge through the practice of farming and the careful observation of their own particular farms and localities. To improve their collective profitability, they could compare their personal knowledge with others and share it among neighboring farmers through local organizations, agricultural fairs, and periodicals. A second view was advocated by so-called gentlemen farmers who distinguished themselves from ordinary farmers and exchanged knowledge in learned agricultural societies. This view held that "scientific farmers" were uniquely qualified to produce agricultural knowledge by relying on their own training in empirical methods combined with practical farming experience. A third view distinguished between and argued for the superiority of the knowledge produced under experimental conditions by scientifically trained professionals over the experiential knowledge produced by farmers in the practice of farming. Ultimately, a melding of the second and third views prevailed.

In alliance with gentlemen farmers who valued scientific training, pressure groups made up of scientists, journalists, and industrialists advocated successfully for high levels of public investment in an agricultural research system guided by professional scientists (Danbom 1986). Over a period of more than fifty years, a series of federal and state laws laid the foundation for the triad system of agricultural science – teaching, research, and extension. In the 1862 Morrill Act, Congress provided land that could be sold by the states to fund land-grant colleges where scientific principles of agriculture would be taught. In that same year, Congress established the U.S. Department of Agriculture (USDA), which initially had few research responsibilities. In 1887 the state agricultural experiment stations were established through passage of the Hatch Act for the purpose of conducting agricultural and rural research in cooperation with the colleges of agriculture. A second Morrill Act was passed in 1890 which allocated more operating money for the colleges and established separate, racially segregated colleges for blacks. The 1914 Smith-Lever Act established the county extension agent system to disseminate the fruits of land-grant research to people in the countryside. Thus only a little more than a century ago a system in which farmers relied primarily on themselves and their neighbors for the knowledge they needed was replaced, and agriculture became a subject of publicly supported, scientific education, investigation, and dissemination in the United States.

Although advocates of the public agricultural research system espoused the powerful rhetoric of agrarianism, furthering the interests of farmers was not their main goal. Rather, their focus was on agricultural *production*, which was viewed as a critical means of invigorating the national economy and moving it toward industrialization (Kirkendall 1987; Marcus 1985). Supporters of the land-grant system assumed that the farmer's lack of scientific knowledge was the major factor limiting agricultural production. Thus, as David Danbom (1986) put it, "the farmer was subtly transformed into a benighted hick, yokel, or rube, who did work requiring brawn but precious little brain." And to replace the self-reliant and independent agrarian image, advocates of the public agricultural research system depicted farmers as people whose problems required the help of scientifically trained professionals.

In contrast, Danbom (1986) showed that most farmers at the turn of the century did not see the development of scientific understanding of agriculture as the solution to their severe economic problems, whose source they believed lay instead with the exploitative tactics of railroads, bankers, farm implement industries, and others. Farmers' discontent found expression in populist organizations that strongly opposed the creation and funding of the agricultural research and extension system. They preferred their own educational programs through which farmers themselves developed and shared knowledge about farming practices and marketing techniques. The research establishment was generally hostile to the populist movements and those who disagreed with the need to professionalize agriculture. Instead, the research institutions made powerful alliances with businessmen, bankers, and highly specialized, commercial farmers who tended to be sympathetic to production-oriented research goals and did not accept populist definitions of agricultural problems.

The tensions between a substantial portion of the farming community and the growing number of land-grant university scientists did inspire certain creative arrangements. For example, at the turn of the century, the "Wisconsin Idea" declared that "the boundaries of the University were the boundaries of the state" (quoted in Stevenson and Klemme 1991:1). At the urging of Wisconsin farm organizations and sympathetic state legislators, institutional arrangements such as farmers' institutes held for one to three days around the state and winter short courses for farmers held at agricultural college campuses provided opportunities for farmers not only to learn from but also to give critical input to university professors. In turn, by interacting with farmers and demonstrating the benefits of scientific and technological modernization, scientists tried to overcome farmers' resistance to accepting their advice.

Successes with these institutional arrangements provided the rationale for federal funding of state extension programs through the 1914 Smith-Lever Act. County extension agents who resided in farm communities would bring new scientific knowledge into the daily lives of rural people. Unfortunately, the input from farmers and citizens envisioned by the Wisconsin Idea, in which "experts were on tap, not on top," never became the norm (quoted in Stevenson and Klemme 1991:1). Instead it slowly faded and was replaced by the perception that the scientific expert knew more than the farmer. John Bennett (1986) argues that ever since, the communication flow has been from the scientist to the practitioner. Meanwhile, the farmer-generated knowledge that had shaped agriculture for thousands of years was slowly hidden from history.

Although the direction in which knowledge appears to flow was established earlier, it was not until the 1940s and 1950s that scientific agricultural research dramatically transformed agriculture in the United States. During that postwar period, a chemical and mechanical revolution in agriculture took place. Bigger, more powerful, and more versatile machines came on the market. A new array of petroleum-based pesticides and fertilizers were developed and disseminated widely. Breeders introduced ever-improving crop varieties and animal breeds. All of this contributed to spectacular increases in production. And the publicly supported system seemed finally to pay off on earlier investments. Fred Buttel (1993a) enumerated the direct consequences enjoyed by the agricultural research institutions: the expansion of budgets, the creation of a class of famers who were attentive to the new technologies, and the establishment of federal agricultural commodity policies to deal with surpluses. Most important, the highly-visible success stories and the rhetoric of the times pledged to eliminate hunger from the United States and even the world, either through the export of agricultural surpluses or the export of technologies to ignite a Green Revolution.

Questioning Agricultural Science

The knowledge that has flowed from the public and private institutions of agricultural science has been one of the most profound forces behind the industrialization of agriculture which the sustainable agriculture movement has found to be so problematic. Buttel (1993a) described how many agricultural scientists saw the advancements in U.S. agricultural technology as unquestionably good and acknowledged no harmful effects flowing from them or at least none that outweighed the benefits during the postwar period. But mainstream agriculture and the research system were not without critics. Most notably, a small but vocal group of ecologically oriented advocates of organic agriculture emerged in the 1940s, including the organiculturalist J. I. Rodale, biodynamicist Ehrenfried Pfeiffer, author and organic farmer Louis Bromfield, and outspoken critic of the moldboard plow Edward Faulkner. Although these and other predecessors of the contemporary sustainable agriculture movement repeatedly challenged the agricultural research establishment, they had little direct impact at the time. During that postwar generation, science became enshrined as the vehicle of progress that was inevitable and good. Consequently, it was not a friendly time for radical social critics, "who came to be viewed as cranks at best and communists at worst" (Danbom 1986:121).

Space for a different attitude developed with the social movements of the 1960s and 1970s. The passions and concerns that stimulated activism in civil rights, antiwar protests, New Left politics, environmentalism, and feminism also energized a critical examination of the scientific and technological foundations of industrial society (Gottlieb 1993; Harding 1986; Winner 1986). The agricultural research system was not unscathed by this tendency to examine how society shapes and is shaped by science and technology. Much of the criticism has been focused on the land-grant system because it is supported by public funds and is therefore thought to be more accountable to the public than corporate research programs. The public agricultural research system has come under intense scrutiny not only from farmers but also from consumers, environmentalists, farmworkers, feminists, and animal welfare advocates. Critics who are largely external to the institutions have been joined by sympathetic members of the research community itself, indicating that agricultural science is not monolithic, inflexible, and hegemonic. A range of critical questions has been raised about ecological, social, and economic consequences of agricultural science and the technologies it has spawned. The most prominent critiques of the crucial role played by the formal institutions of agricultural research and dissemination seem to revolve around two central knowledge questions.

What Are the Worldview Assumptions Underlying Agricultural Science?
In his now-famous collection of essays titled *A Sand County Almanac*, Aldo Leopold recognized that a dichotomy exists in agriculture as well as in other forms of land use. He described a "cleavage" between "man the conqueror *versus* man the biotic citizen; science the sharpener of his sword *versus* science the searchlight of his universe; land the slave and servant *versus* land the collective organism" (Leopold [1949] 1970:260–61). This distinction refers to what might be called a mechanistic, reductionistic view of the world on the one hand and a more holistic, organic view on the other. These competing worldviews provide the basis for one of the most fundamental critiques of the classic view of science in general and of agricultural science in particular.

The classic view of science, articulated originally by Francis Bacon and René Descartes in the seventeenth century, holds that only scientists' unbiased observation and detached, logical reasoning can produce and establish accurate accounts of the natural world and its (assumed) underlying uniformities and unchanging "laws." Bacon used very explicit, controlling imagery to articulate that the purpose of uncovering those laws should be to dominate nature for human advantage (Leiss 1972; Merchant 1980). Bacon thought that if a scientific community worked together as a group and was supported by the state, scientists could bend nature to their will and further the material welfare of humanity.

While Baconian science rested on the domination of nature for human benefit, Descartes provided the metaphorical image of nature as machine. Descartes held that if scientists can discover nature's knowable elements that can be generalized and remain unchanged across space and time, nature can be made orderly – like a well-functioning machine – and thus be controlled. Accordingly, scientists acquire objective knowledge about nature by using standardized conditions and systematic measurement to test general theories and hypotheses. To correct for what are perceived as potentially contaminating factors – an observer's values or biases – knowledge claims are verifiable with rigorous enforcement of methodological standards and the ability to replicate experiments regardless of spatial location.

The atomistic, reductionistic view of the world remains at the core of science. The application of science to agriculture illustrates how mechanistic thinking and the doctrine of domination of nature have worked (Cobb 1984; Crouch 1995; Kirschenmann 1992a). Since its origins, agricultural science has focused narrowly on maximizing production from the agricultural landscape. As Buttel (1993a) pointed out, productionism emphasizes that increased yields and higher labor efficiency are intrinsically socially desirable and that increased production is symbolized by the "magic bullet." For example, dairy scientists have created recombinant bovine growth hormone (rbGH) to extend the lactation of the treated dairy cow so she will produce more milk. Plant breeders have perfected tomato varieties that are hard enough to survive the grip of the mechanical harvester. Biotechnologists are working to change the genetic composition of the corn plant to make it more resistant to damage from the herbicides prescribed for use in corn production. These technological innovations embody "progress," which is evaluated solely in terms of output, utility, and efficiency. What is obscured is the entire economic, social, physical, and ethical context in which food is produced and in which human and other beings live. In ignoring the broader context, as Stan Rowe (1990:129) put it, the university "has become a know-how institution when it ought to be a know-why institution."

Underlying the answers to the "know-how" questions is an attempt to control all aspects of the growing environment and to create uniformity on the agricultural landscape. Agricultural science, like all science, tries to generate universal truths that hold regardless of social or spatial location. To do this, scientifically acceptable experiments must be conducted in artificial environments with a reduced number of variables and must be detached from local contexts so that they can be replicated elsewhere (Suppe 1987). The knowledge generated and the technologies produced are generalized solutions to agricultural problems that are supposed to work in the same way anywhere and everywhere. But Frederick Suppe (1987) has shown that problems arise when farmers must apply that general knowledge in dynamic and unique natural and social environments. In all its variability, nature is diminished as a factor in conventional agricultural production to the greatest extent possible. And the farmer is diminished too. Cornelia Flora (1992:93) summarized the effect: "Farmers are becoming interchangeable, as knowledge of the local conditions is less relevant than following the best management practices and package directions for chemical and fertilizer applications."

According to Carolyn Merchant (1980; 1992), a profoundly different way of conceptualizing the environment prevailed before the scientific revolution. In what she characterized as "organic society," the Earth was considered to be a living organism, an image that served as a cultural constraint and restricted the actions of human beings. Elements of the organismic perspective have emerged in the contemporary sustainable agriculture movement. At the root of many alternative conceptions of agriculture is a whole-systems perspective, an appreciation of the interconnectedness and interdependencies of all life (Freudenberger 1986). For instance, a farm should be recognized not as a mere compilation of unrelated parts but rather as an integrated system in which the various elements interact with each other, with the environment, and with people. William Lockeretz and Molly Anderson (1993:12) explained that this was the connotation originally intended by the word *organic* in the term *organic farming*.

Sir Albert Howard, a British scientist who contributed much to the early organic farming movement in the 1940s and 1950s, clearly advocated this holistic view. By insisting on the importance of soil flora and fauna, Howard stressed that soil is alive, not a dead, inert, material substance to be manipulated. Howard thought that most agricultural scientists missed this insight because they distanced themselves from the field and became "laboratory hermits" (quoted in Peters 1979:255). Howard (1945:vii) believed that scientific knowledge was becoming more and more fractionalized: "The experts, as their studies become concentrated on smaller and smaller fragments, soon find themselves wasting their lives in learning more and more about less and less. . . . Everywhere knowledge increases at the expense of understanding." The understanding that the experts miss is of the wholeness of nature, what Howard called "The Wheel of Life."

The lack of a holistic perspective is reflected in the institutional structure of agricultural science. Internally, agricultural science is sharply divided into disciplines such as agronomy, agricultural economics, rural sociology, animal sciences, and nutrition. In turn, agricultural scientists are trained to answer questions within narrow specialty areas, often focused on a particular commodity. Each discipline maintains its own research orientation, criteria for choosing research problems, language and communcation patterns, and publication standards. These mechanisms strongly affect the choices that scientists make about the research questions they ask by affecting their prospects for publication, tenure, promotion, and research funding. Lawrence Busch and William Lacy (1983) have shown that public sector agricultural scientists tend to ignore questions that are outside their particular specialty and often regard such questions as unimportant. Consequently, the process of conducting research and disseminating results is fragmented with little collaboration across disciplinary lines. Such fragmentation precludes analysis of complex ecological and social systems and minimizes the prospects for resolution of environmental and social problems.

To overcome these obstacles, some sustainable agriculturalists have advocated a more systems-oriented, interdisciplinary approach to agricultural science that would try to reflect the integrated realities of agricultural systems. In part, this means applying the science of ecology to agriculture, creating what is known as agroecology. The science of ecology assumes that an organized reality exists in nature, a reality that is fundamentally relational with humans embedded within, rather than separate from, nature. Despite the growing interest in ecology and systems-level analysis, this area has received little attention from the land-grant colleges (Lacy 1993).

Knowledge for sustainability emphasizes the need to understand the fundamentally relational character of the world and challenges humans to consider how we can preserve the integrity and stability of the natural systems within which agriculture occurs (Kirschenmann 1992a). The worldview associated with organic or sustainable agriculture recognizes that human beings can act in the world in either a destructive or constructive manner. In sharp contrast is the worldview of agricultural science that assumes the purpose of knowing the world scientifically is to dominate nature for the benefit of humans, without giving much thought to the consequences such an approach may have on the natural system as a whole.

Whose Interests Has Public Agricultural Science Served?

From the classic view of science, strict adherence to scientific method can detach the perceptual, cultural, and physical attributes of the knower (i.e., the subjective) from the knowledge produced. Societal influences are thought to impinge on scientific activity only to the extent that nonscientific criteria are employed in the generation of theories or in the allocation of research funds. Thus, according to Baconian science, scientists must operate in institutions that are free from external manipulation so as to remain independent of social and political pressures. While autonomy has been the basis for claiming the objectivity and thus the superiority of scientific knowledge, powerful critques in recent decades have challenged the assertions of neutrality and independence of public agricultural science.

Some of the most incisive questions about neutrality were first raised by Rachel Carson. Her *Silent Spring* (1962) was an accessible and powerful disclosure of the hazards of pesticides and began a long debate about the effect of agricultural chemicals on human health and the environment. But *Silent Spring* was more than that. Carson provided a strong critique of how science had been used both to promote synthetic pesticides and to ignore and hide their ill effects, all to the benefit of the chemical industry. Robert Gottlieb (1993), in his history of the environmental movement, explained that Carson was opposed to the idea of a scientific elite and to the fact that an elite science could be purchased by industry and thus could be corrupted. Instead, Carson thought that expertise ought to be democratically grounded in the fabric of everyday life. She insisted that pesticides were a public rather than a technical issue because their environmental consequences threaten not only to silence birds in the springtime but also to endanger human life. Thus Carson had a broad view of science: "We live in a scientific age; yet we assume that knowledge is the prerogative of only a small number of human beings, isolated and priestlike in their laboratories. This is not true. The materials of science are the materials of life itself. Science is part of the reality of living; it is the what, the how, and the why in everything in our experience" (quoted in Gottlieb 1993:82).

Although Carson's critique was general, exposing the undemocratic characteristics of the public land-grant college complex in particular was the point behind Jim Hightower's *Hard Tomatoes, Hard Times*. As director of the Agribusiness Accountability Project, Hightower first published his fiery report in 1972 with the spirit of earlier populist agitators. He argued that the land-grant colleges, experiment stations, and extension services have been the scientific and intellectual foundation of "a social and economic upheaval in the American countryside . . . a protracted, violent revolution" (Hightower 1976:87). According to Hightower, the land-grant college complex has worked to the benefit of agribusiness corporations and large-scale farmers and to the detriment of small farmers, black farmers, sharecroppers, farmworkers, consumers, and the environment. Such a close relationship had been established between industrial farming interests and agricultural research institutions that Hightower (1976:102) asked rhetorically: "Where does the corporation end and the land grant college begin?" Hightower (1976:98) called on agricultural research – as a publicly funded enterprise – to serve broader interests and to turn away "from the erroneous assumption that big is good, that what serves Ralston Purina serves rural America. It means research for the consumer rather than for the processor. In short, it means putting the research focus on people first – not as a trickled-down afterthought."

Other radical critics echoed this call. Most notably, farmer-author Wendell Berry included an eloquent critique of the land-grant system in his widely read book *The Unsettling of America: Culture and Agriculture*, published in 1977. Berry (1977:165) was scathing in his criticism of the land-grant universities, which he claimed betrayed the public trust because their work has served the most powerful:

Lacking any moral force or vision of its own, the "objective" expertise of the agricultural specialist points like a compass needle toward the greater good of the "agribusiness" corporations. The objectivity of the laboratory functions in the world as indifference; knowledge without responsibility is merchandise, and greed provides its applications. Far from developing and improving the rural home and rural life, the land-grant colleges have blindly followed the drift of virtually the whole population away from home, blindly documenting or "serving" the consequent disorder and blindly rationalizing this disorder as "progress" or "miraculous development."

Berry maintained that in encouraging farmers to become more professional and efficient, agricultural scientists have made agriculture more of a business than a way of life. In the process, farming has lost the special value that the agrarian ideal had attributed to it.

While leaders in the research system have often claimed that they are preserving agrarian values, critics argue that the consequences of much agricultural research have undermined those values in ways that for too long went unanalyzed from within. Nevertheless, the ecological and agrarian criticism generated by Carson, Hightower, Berry, and others external to the public agricultural research system has inspired some social scientists in recent years to examine agricultural science more critically than they had previously. Their studies have demonstrated that agribusiness and large-scale industrial farmers have had a disproportionate influence over agricultural science (see Buttel, Larson, and Gillespie 1990). A fairly narrow set of social interests have influenced the research questions that public agricultural science has (and has not) asked and the way answers to those questions are expressed in technologies (Friedland and Kappel 1979; Kloppenburg 1988). Those technologies have tended to replace production activities once done on the farm with inputs produced by and purchased from industrial capital off the farm. As a result, farmers need more and more capital. And the continuous development of innovations fuels what Willard Cochrane (1979) dubbed a "technological treadmill" in which farmers who adopt new technologies enjoy benefits as early innovators; as the adoption spreads, however, increases in production cause prices to fall, and a new round of innovations is required for the farmer to get ahead.

Operating under a productionist ideology, the public agricultural research system defends its research priorities by pointing to the enormous increases in output its work has facilitated. Agricultural research has also contributed to the creation of such conveniences as the availability of out-of-season produce and the increased shelf life of food products. Many social scientists, as well as sustainable agriculturalists, have argued, however, that despite these and other perceived benefits of the existing system, most agricultural scientists have acted as what William Friedland (1978) called "social sleepwalkers" largely ignoring the external social and environmental costs associated with the technologies that now characterize industrial farming.

In focusing on the questions of interest to particular disciplines and a narrowly defined clientele, scientists have not asked research questions that are in the broader public interest. Small farmers are not the only ones that have been excluded. According to William Lacy's (1993) analysis, there is a general tendency to neglect those who are separate from people in research institutions whether by physical space, socioeconomic status, time (future generations), gender, or ethnicity. And Lacy (1993:42) fears that "the broad experiences and innovative ideas required to make research programs responsive to a sustainable agriculture agenda may be lacking."

Indeed, there is an astounding lack of diversity in the agricultural research community. The institutions of agricultural science are made up predominantly of white males. Women and ethnic minorities are underrepresented and receive fewer benefits from the system. For example, Busch and Lacy (1983) found that as of the late 1970s only 1 percent of agricultural scientists at public institutions were at the seventeen predominantly black, or 1890, institutions. Allocations from the U.S. Department of Agriculture to these black land-grant colleges have been relatively small in comparison to those given to other colleges, constituting a form of "institutional racism" (Hightower 1976:105). Joel Schor (1992) argued that these separate and poorly funded institutions have found it extremely difficult to participate as full partners in the system. Black farmers have been leaving agriculture at over twice the rate of white farmers during the past fifty years, despite recent programmatic efforts to reverse historical discrimination (Schor 1996).

One reason may be that the research institutions that have developed the dominant agricultural technologies are overwhelmingly staffed by whites and have not applied their research efforts to issues relevant to keeping black farmers on the land.

The number of women in agricultural science is significantly lower than in other fields of science. Busch and Lacy (1983) found that in the late 1970s only 4 percent of agricultural scientists were women. Most of these women are in stereotypically female fields such as nutrition and home economics. Reflecting a general tendency in society, the rise of agricultural science at the turn of the century was parallelled by a rise in so-called domestic sciences. Reinforced and promoted through educational texts, the popular media, and religion, a domestic ideology established a set of behavioral norms that defined the ideal role of women in the context of the household, not the fields (Garkovich and Bokemeier 1988).

According to Katherine Jellison (1993), farm women in the midwestern states often favored the adoption of modern field and household technologies during the first half of the century. But they resisted the idea that such technologies should release them from their role as productive actors on the farm; instead, in their vision of agriculture women's roles were central. Nevertheless, within the land-grant college complex, a division developed between the scientific knowledge about agricultural production that was distributed to male farmers and the scientific knowledge about domestic concerns that was principally developed by and taught to women. Carolyn Sachs (1983) argued that the colleges of agriculture institutionalized and solidified a sexual division of labor on the farm. Moreover, according to Jane Knowles (1985:54), these two spheres have not been treated equally: "Over the years, the separation between the two has been very strict, with disproportionate staff and funding resources being devoted to the technical [agricultural practices] delivery system."

Putting Public Agricultural Science to Work for Sustainability

Clearly, knowledge issues are critical in strategizing about how sustainability goals might be achieved. Accordingly, some individuals and organizations in the sustainable agriculture movement have worked long and hard to put science to work for sustainability, and these reform efforts have met with some success during the last fifteen years. The first policy milestone was the 1980 U.S. Department of Agriculture's *Report and Recommendations on Organic Farming*, which was meant to educate policy makers and the scientific community about organic agricultural systems. Secretary of Agriculture Robert Bergland commissioned the study for two reasons: First, the USDA had difficulty answering the numerous requests it received for information on organic agriculture. Second, Bergland sought to identify research and educational programs that would contribute to a better understanding of organic farming systems which he hoped might address some of the detrimental aspects of conventional agriculture (Harwood 1993). Not surprisingly, one of the major recommendations of the report was that integrated, multidisciplinary research projects examining organic farming be established.

Armed with the USDA report, proponents of low-chemical production were able to get bills introduced in Congress in 1982. The purpose of the legislation was to develop scientific knowledge about organic farming and to disseminate that knowledge to family farmers. Those bills were defeated. As Garth Youngberg and colleagues (1993:298) point out, the bills' proponents "had seriously underestimated the negative symbolism of organic farming, which had long since been dismissed by conventional agriculture as little more than a primitive, backward, nonproductive, unscientific technology suitable only for the nostalgic and disaffected back-to-the-landers of the 1970s." Following the defeat, advocates of organic farming and other low-input approaches began to embrace the term *sustainability* in an effort to diffuse opposition in the political arena.

This rhetorical shift in language, as well as increased lobbying activity around agricultural issues by both national organizations and grassroots groups, led to further policy gains. The basic concepts of the 1982 legislation were eventually authorized in the 1985 farm bill, which created the Low-Input Sustainable Agriculture program (later renamed the Sustainable Agriculture Research and Education program, SARE). The program provides financial support for new agricultural research and education projects that will increase the profits and reduce the risks of farming methods that require substantially lower than conventional levels of synthetic chemicals and fertilizers. SARE has provided funding for hundreds of projects proposed by a wide range of public and private organizations, including farmers and private research and education groups. One reason for the focus on increasing the availability of government research funds for sustainable agriculture is that corporations – which have been the source of much agricultural research funding – have little incentive to do research into farming practices that might reduce inputs and thus lower profits.

More legislative victories occurred in the late 1980s and in the 1990 farm bill. First, the Appropriate Technology Transfer for Rural Areas (ATTRA) program was established in 1987 (with funding from the U.S. Fish and Wildlife Service) to disseminate information on alternatives to conventional practices. Ten years later ATTRA's staff had responded to over one hundred thousand requests from farmers, county extension agents, and others seeking free information and technical assistance on low-input and sustainable agricultural practices (ATTRA 1998). Another milestone came in the 1990 farm bill, which required that any research projects receiving funding assistance through SARE must involve farmers. In addition, the Sustainable Agriculture Technology Development and Transfer program was created to educate and train extension agents in alternative agriculture.

The social momentum inspiring these policy actions and the provision of funding for applied research and education in sustainable farming systems have sparked new interest among researchers and inspired the creation of new institutional arrangements and centers for sustainable agriculture at several land-grant universities. For example, a grassroots-based, rural advocacy organization, the Wisconsin Rural Development Center (WRDC), was instrumental in a campaign to establish the Center for Integrated Agricultural Systems (CIAS) at the University of Wisconsin–Madison in 1989. The CIAS was created to address the concern that for too long the land-grant system has neglected the needs of and input from farmers, particularly those with small and medium-sized operations and those who are developing alternative farming systems (Stevenson and Klemme 1991). Other prominent institutions are the Leopold Center for Sustainable Agriculture at Iowa State University, the Sustainable Agriculture Research and Education Program at the University of California–Davis, the Center for Sustaining Agriculture and Natural Resources at Washington State University, and the Center for Sustainable Agriculture Systems at the University of Nebraska.

The transformation of contemporary agricultural research institutions requires more than redirecting funds for a new research agenda and creating new institutions with an expanded constituency base. The debate revolves not only around *what* research is funded, but also *how* that research is undertaken. Thus farmers, activists, and academics have advocated and experimented with new, alternative models for how agricultural research is conducted and who participates in the process (Bezdicek and DePhelps 1994; Taylor 1990; Thornley 1990).

William Lockeretz and Molly Anderson (1993:13–16) identified and evaluated five agricultural research alternatives that are being explored: (1) developing new forms of multidisciplinary and interdisicplinary research teams to transcend the disciplinary boundaries characteristic of reductionist science; (2) conducting research grounded in agroecological principles that recognize the holistic interconnections between the human/cultural and the biophysical elements of farming systems; (3) recognizing that the development of new systems will require replacing technological inputs with farmers' understanding, information, and management as critical inputs into the production process; (4) designing, carrying out, and evaluating more research on working farms where a host of factors come together and which present more realistic experimental conditions than laboratory and field station research; and (5) increasing the involvement of farmers in the research process (e.g., on advisory councils, in focus groups, as research partners) so as to gain a better understanding of the values and needs underlying various production and marketing options. To this list, a farmer, Frederick Kirschenmann (n.d.:9), added the idea that new relationships with extension agents ought to be established to improve and facilitate "information exchange, rather than technology transfer" from scientist to practitioner.

Among the organizations that are working to transform agricultural science are regional networks around the country called Sustainable Agriculture Working Groups (SAWGs). For example, in 1988 the Midwest SAWG was developed to give its twenty-seven member organizations concerned with sustainable agriculture a more coordinated voice on federal policy issues. The Midwest SAWG's agenda for the 1990s focuses on efforts to "seek full funding for sustainable agriculture research and redirect all publicly funded research toward sustainable goals" (SAWG n.d.). Like their counterparts in the South, Northeast, and West, the Midwest SAWG links local concerns and efforts at building sustainable agriculture to the achievement of changes in national research policy.

At the national level, one of the most significant organizations advocating "putting science to work" for sustainability is the Henry A. Wallace Institute for Alternative Agriculture established in 1982 (Youngberg, Schaller, and Merrigan 1993:311). In contrast to the grassroots base on which many of the member organizations of the SAWGs rely, the Henry A. Wallace Institute (1992:6) is highly expertise-oriented and describes its role as "intellectual advocate." The professional staff promote federal policies that encourage the adoption of sustainable farming practices and try to increase the understanding of and support for sustainable agriculture in agricultural science institutions.

The emphasis placed on transforming the institutions of agricultural research and extension is certainly understandable. After all, modern science has been very effective in answering certain questions, and it is logical to suggest that it could contribute to the growth of sustainable agriculture. Tracy Hewitt and Katherine Smith (1995:1), authors of a report released by the Wallace Institute, expressed this confidence: "If modern science can map the human genome, send a satellite to Jupiter, and transfer genes from animals to plants, then certainly we can find ways of increasing crop yields with fewer chemicals and other environmentally threatening inputs and practices." Youngberg and colleagues (1993:311) at the Wallace Institute argued that "the problem is not science. It is that science has not been developed and used effectively by scientists to ensure the sustainability of agriculture." From this view, it is primarily scientists, not farmers, who should determine the "facts" that will help to create a new agriculture.

Several analysts of sustainable agriculture have raised questions about this uncritical call for science to bolster the ideas articulated and the practices used by advocates of sustainable agriculture. For example, Frederick Buttel (1993b:31) pointed out that the focus on "influencing science and research policy ironically reinforces the privileged role of science in shaping agriculture." Similarly, Patricia Allen and Carolyn Sachs (1993:155–56) referred to this tendency as a "reification" of natural science that fails to question a basic premise of existing science that humans can dominate and master nature through knowledge. Allen and Sachs (1993:156) were also concerned that there has been an overemphasis on natural rather than social sciences and that "there has been little serious investigation of the social, political, and economic relations that are needed to encourage sustainable agriculture." In other words, research and technical innovations alone will not achieve the social goals in sustainable agriculture. Frederick Buttel and Gilbert Gillespie (1988) have warned that the increased respectability of sustainable agricultural initiatives at agricultural research institutions may be similar to other "development fads" such as farming systems research, women in development, and basic needs. These and other progressive symbols were originally nurtured by social movement organizations, but the least threatening versions have been adopted by dominant institutions without effecting substantial change in the status quo. Thus without a constructive skepticism of public research institutions, sustainable agriculturalists may find that the goals of the movement are diluted and that ultimately the movement is weakened greatly or demobilized completely.

Creating an Alternative Knowledge System

Academics and professional activists attempting to reform agricultural science policy represent just one arm of the sustainable agriculture movement. The emphasis on reforming science to better serve sustainable agriculture tends to obscure the less visible but more pervasive activities of farmers who are now engaged in producing and reproducing a landscape of sustainable alternatives independent of what research gets done in the land-grant universities and whatever the latest farm bill does or does not say. As they face the paucity of information about practices that deviate from main-stream agriculture, farmers using unconventional methods have tried to de-velop an alternative knowledge system that unearths and builds on ideas and techniques ignored or marginalized by agricultural science. This strategy draws on the problem-solving, creative capacities of farmers acting both in-dividually and collectively.

My central interest in the chapters that follow is to explore how this pro-cess of creating and disseminating alternative knowledge works in the sus-tainable agriculture movement as it originates and functions *outside* of the formal institutions of agricultural research and as it operates at the *local* level among farmers who endeavor to create alternative farming systems and to share their knowledge with each other. Four primary factors moti-vated my study of this alternative knowledge system: recent organizational activity among sustainable farmers, recent theoretical interest in farmers' local knowledge, the knowledge dimension in social movement theory, and the strategic importance of local approaches to social change.

Recent Organizational Activity Among Sustainable Agriculturalists
In recent years, sustainable agriculturists have created a variety of organiza-tions for the specific purpose of generating and exchanging knowledge use-ful to alternative farmers. Today a host of rural organizations and farmer networks emphasize farmer-generated knowledge, promote holistic and ecological thinking, and embrace practical research (Bird, Bultena, and Gardner 1995; Enshayan, Stinner, and Stinner 1992). These organizations transmit information and ideas about sustainable agriculture through such channels as annual meetings, field days, conferences, collective marketing strategies, books, magazines, and videos. While these organizations share common elements, a range of approaches constitute the alternative knowl-edge system in the sustainable agriculture movement.

Until the mid-1980s, during the many years of neglect by the dominant agricultural research and extension system and in the face of what was often outright hostility to unconventional ideas, advocates of agricultural alterna-tives had little choice but to create their own research institutes. Some of these earlier organizations have long conducted experimental research into organic and other alternative farming methods. While many generate knowledge based on the work of their own staff scientists, their leaders have also recognized the value of farmers' knowledge and developed a politi-cized vision of research. Although J. I. Rodale, founder of the Rodale Insti-tute in 1939, applauded sympathetic scientists as a "maverick" breed, he encouraged farmers to rely on their own "common sense" and "unscientific gumption'' (quoted in Peters 1979:262). Similarly, in founding the New Alchemy Institute in 1968, the young scientist John Todd (1971:55) sought to create ''a non-destructive science based upon an ecological ethic, a science which by its very nature cannot be abused for power or unneeded gain. This science and its technology should be the offspring, not of a few men in remote laboratories, but of all of us working as part-time researchers in our yards, barns and garages. It will be our own and we will grow with it.''

Today, among the most prominent of such research organizations is the Land Institute in Salina, Kansas, founded by Dana and Wes Jackson in 1976. Ecologists at the Land Institute study the native prairie as an ''analogy'' for the development of perennial polycultures that might potentially substitute for annual monocultures (Jackson 1990). Although he is very interested in creating a formal agricultural science based on an ecological understanding of whole systems, Jackson (1987:14) also has great respect for what he calls ''folk knowledge,'' which is ''accumulated through tradition, daily experience, and stories, mostly in an informal setting.'' Jackson (1987:16) deplores the loss of this ''cultural information'' and worries about the difficulty in regaining it given the low ''eyes-to-acres ratio'' that results when farmers are driven from the land as a consequence of the industrialization of agriculture.

Often founded by scientists themselves or by advocates who employ scientists, these research institutes reflect a vision of a new, more socially responsible science that has its origins in progressive social movement critiques of science. This vision has focused less on transforming the formal institutions of scientific research and more on creating new organizational principles guiding the creation of knowledge for an alternative agriculture. Primary among these principles is a commitment to the value of farmer-generated knowledge, as well as a desire to create a nonelitist, more democratic system of generating and disseminating knowledge. In other words, a key feature of the alternative knowledge system is a recognition that knowledge is not only the purview of a relatively small number of people in universities or laboratories but a human capacity that can be developed and enriched in everyday life.

Perhaps the greatest potential for realization of this democratic view of knowledge lies not so much with the alternative research institutes as with local sustainable agriculture organizations that are primarily farmer-based. Often with private, nonprofit advocacy organizations providing the impetus, the number of local groups has exploded across the United States during the last decade. For example, the Alternative Energy Resources Organization has been instrumental in organizing numerous farm improvement clubs made up of farmers and ranchers in Montana and Idaho (Matheson 1997). The Land Stewardship Project in Minnesota has organized several sustainable farming associations. Other groups include the Arkansas-Oklahoma Sustainable Agriculture Network, the Maine Organic Farmers and Gardeners Association, the Northeast Organic Farming Association, the Michigan Agricultural Stewardship Association, and the Ohio Ecological Food and Farm Association. This list is not meant to suggest that these organizations share identical aims and activities; the point is that there are a great many local and statewide farmer-based groups engaged in a range of activities, the primary purpose of which is to generate and exchange knowledge.

At one end of the spectrum are organizations that focus almost exclusively on conducting their own on-farm, experimental trials into alternative farming methods, often with assistance from scientists at the land-grant universities. A well-known example of this approach is Practical Farmers of Iowa (PFI), which was founded in 1985 and has about 450 members. According to one of the group's leaders, Ronald Rosmann (1994:34), its purpose is "to provide farmers with information about environmentally sound, lower cost, profitable farming techniques." Rosmann explained that the group members chose the word *practical* for its name because they wanted "to be realistic" and because, to "appeal to as many farmers as possible, words such as 'organic,' 'biological,' and 'regenerative' were avoided." The organization has had a close relationship with Iowa State University, but its founders felt that for research to be sufficiently practical it had to be conducted on farms with farmers as full partners. Its farmer-initiated, on-farm research trials allow these farmer researchers to do statistically valid comparisons of competing techniques such as different weed control methods. In this way, PFI has tried to meld what it perceived to be two extremes in previous approaches to agricultural research: one based on farmers' unreplicated field observations and the other based on small, unrepresentative plots done off the farm on university field stations. Farmer research organizations such as PFI reflect an attempt to bridge the gap between farmer-generated knowledge and systematized knowledge derived from scientific methods.

Closely related to farmer organizations that actively conduct experimental research as a group, participants in the sustainable agriculture movement have also created what are often called farmer-to-farmer networks. Network activities enable farmers to share knowledge that they have generated through experience rather than through systematic experimentation. These organizations can be distinguished from casual networking that occurs among neighbors and friends who discuss ideas and practices over the fence or in a supply store. Farmer-to-farmer networks are key organizational elements of an alternative knowledge system that draws on farmers' personal knowledge and extends that knowledge by developing deliberate mechanisms for sharing ideas, innovations, and techniques among a wider community. Given the increasing amount of organizational activity around the creation and exchange of knowledge at the local level and the lack of analysis of farmer networks, I felt that it was important to examine the role of these networks in the sustainable agriculture movement.

Theoretical Resurrection of Farmers' Local Knowledge

The increased level of local and statewide organizational activity for the explicit purpose of generating and exchanging knowledge useful to alternative farmers is complemented by theoretical efforts to resurrect farmer-generated knowledge that was hidden from history with the rise of systematized, formalized agricultural science. The value of farmer or "indigenous" knowledge has been recognized in the field of international agricultural development for some time (Brokensha 1989; Chambers 1983; McCorkle 1989; Thrupp 1989). More recently, some analysts of sustainable agriculture have appreciated farmers as producers of knowledge as well as of agricultural commodities. For example, reflecting on organic farming practices in the United States, Richard Harwood (1993:150) described "a truly indigenous knowledge system . . . [in which] practitioners assimilate knowledge from each other and from both historical and scientific sources." John Gerber (1992:119) suggested that for farmers "an intuitive understanding of relationships among multiple variables, their confidence in their own observations, and the apparent success of practical solutions produce experiential knowledge that may have more immediate utility than scientific knowledge." In this way, these and other analysts reverse the long-held but faulty assumption that farmers are only recipients not generators of knowledge.

Charles Francis (1990) argued that both how farmers use information and their management ability are thought to be particularly important in farming systems that rely on practices associated with sustainable agriculture. Of course, conventional agriculturalists also use a considerable range of information in managing the technical, financial, and marketing elements of their operations. But conventional reliance on technological inputs may obviate the need for some of the judgments and precautionary measures that alternative farmers use as they try to make decisions on an integrated, whole-farm basis and as they substitute ecological understanding for technological control of growing conditions. To put it a different way, using the biological principles and interactions inherent in agricultural systems requires more complex management and a more nuanced understanding of how that system and its components interact. Developing the ability to do that successfully produces knowledge crucial to alternative farmers, and its dissemination is particularly important because alternative farming techniques and marketing strategies have been ignored by established sources of advice.

To reconstruct agriculture for sustainability, Jack Kloppenburg (1991, 1992) has forcefully argued for the need to draw on farmers' "local knowledge." By local knowledge, Kloppenburg (1991:528) meant the practical, sensuous, personal skill that develops with careful attention to the distinctive yet dynamic social and physical features of a specific locality and that is fundamentally tied to direct experience of a particular place or activity. In developing the idea that local knowledge can be a critical component for the reconstruction of agriculture, Kloppenburg's definition reflected a customary distinction. On the one hand is systematized, specialized knowledge (what is usually referred to as "science"), and on the other hand is a constellation of terms often meant to indicate uncodified knowledge (knowledge that is usually excluded from the definition of science) such as "everyday knowledge" (Mulkay 1979), "tacit knowledge" (Polanyi 1966), "science of the concrete" (Lévi-Strauss 1962), "craft knowledge" (Braverman 1974), "indigenous knowledge" (Warren 1994), "working knowledge" (Harper 1987), and "situated knowledge" (Haraway 1988). Although this distinction between scientific knowledge and local knowledge may not be clear-cut, it has persisted.

To articulate the promising contribution that local knowledge might make to sustainable agriculture, Kloppenburg drew on key insights generated by the new sociologists of science and by feminist critics of existing science. During the last two decades, "social constructivism" has emerged as a school of thought within the sociology of science. Social constructivists have argued that scientific knowledge claims are created and do not necessarily constitute an objective description of the natural world (Knorr-Cetina 1984; Latour and Woolgar 1986; Mulkay 1979). In other words, what we call scientific knowledge may not be any less susceptible to social influences than any other way of knowing. From this position, Kloppenburg (1991:525 n.3) saw "the creation of space for consideration of competing modes of knowledge production, which themselves represent partial understandings."

In articulating these competing modes of knowledge production, Kloppenburg drew on the work of feminist critics of science (e.g., Haraway 1988; Harding 1986; Smith 1987). A central theme in those works affirms the legitimacy of personal experience as a source of knowledge: "Women's distinctive historical experiences – of their bodies, of oppression, of caring (about and for) – makes possible alternative ways of thinking about nature and knowing the natural world'' (Kloppenburg 1991:526). These alternative ways have direct implications for evaluating knowledge claims. Stereotypical science has repudiated the resources of intuition, feeling, caring, sensuous activity, connectedness to nature and others, relatedness, and emotion that have been associated with women and femininity (Fee 1986; Keller 1987). Instead, those attributes have often been associated with nonscientific and local knowledge.

In advocating local knowledge and what Hilary Rose (1983:90) called ''the integration of hand, brain *and* heart,'' Kloppenburg's central point was not simply to replace scientific knowledge with local knowledge. Rather, he wanted recognition that ''science as we have known it has no unique or exclusive claims to truth. There are multiple ways of knowing the world, and we need to embrace the full range of potentials contained in that diversity. Our task is not to choose a single best science from among these variously situated knowledges but to acknowledge that each may be useful for different objectives'' (Kloppenburg 1991:102).

The power and the promise of local knowledge for building a more sustainable agriculture are derived from several key elements. First, as feminists have emphasized, an individual's personal experience, intuition, feelings, and sensual perception are legitimate sources of knowledge which are complementary to the generalized knowledge of science. Second, the context or locality in which knowledge is produced and applied is critical because knowledge is mediated by a host of factors related to an individual's positioning in a particular social and physical place. Third, a farmer's local knowledge develops slowly over time and in relationship to a particular farm; therefore, that relationship must be an enduring one because only over time can the knowledge necessary for sustainable agriculture be developed. This is akin to what David Orr (1996) has called ''slow knowledge,'' which accumulates in a place incrementally through cultural maturation, in contrast to the high velocity at which scientific or ''fast'' knowledge is accumulated and applied today. True wisdom, Orr observes, takes time. Fourth, because local knowledge is thought to be profoundly personal, it is often described as tacit, that is, so tightly bound to an individual (or a cultural group) that it is difficult to communicate to others who have not directly experienced it. As a result, local knowledge lacks the universality associated with scientific knowledge; however, Kloppenburg emphasized that local knowledge is valuable for its practical ability to solve problems at the whole-farm level where farmers must continually evaluate a range of economic, social, and environmental variables as they make decisions in a particular place.

Discussions of local knowledge to date have not been explicit enough about its social character in at least two ways. First, Shelley Feldman and Rick Welsh (1995) pointed out that the concept of local knowledge must incorporate the social diversity that inevitably exists in a given locality. In other words, the emphasis on the knower's unique physical place must not obscure the importance of the particular social location of the knower and how that influences the creation of knowledge. Feldman and Welsh argued that because men and women tend to do different kinds of work on farms and have different lived experiences in a gendered society, there are multiple and partial perspectives from which local knowledge for sustainability might be generated.

Second, treatments of local knowledge have tended to emphasize the deeply personal character of this way of knowing and to ignore the social exchange of such knowledge. Farmer-author Gene Logsdon (1984:4) eloquently describes "taking time to just stand and watch" as the foundation of what he calls "traditional agriculture." Kloppenburg (1991:531) theorized that the generation of local knowledge is a process that results in the creation of information so tightly bound to place that it is applied only with difficulty anywhere else. Paul Richards (1993:67) focused on idiosyncratic "performance," the knowledge-producing activities of "a specific farmer on a specific piece of land in a particular year." In apparent contrast to these treatments, however, the farmer-to-farmer organizational activity occurring around the country suggests that alternative agriculturalists are exchanging their own personal, local knowledge among a wider community.

Although the importance of local knowledge in sustainable agriculture is beginning to be recognized at a theoretical level, little work has been done to explore these insights in concrete instances. Therefore, one of the principal reasons I undertook this study was to examine local knowledge in the context of particular networks of farmers. In so doing, I wanted to be attentive to the possibility that such personal knowledge is being exchanged within these networks and also to the possibility that both physical place and social location shape the production and exchange of local knowledge.

The Knowledge Dimension in Social Movements
Very recent contributions to social movement theory suggest that movements function, in part, as social laboratories where people experiment with creating their own knowledge in the form of new practices, ideas, and orga-

nizational principles (Downey 1986; Eyerman and Jamison 1991; Fals Borda 1988; Wainwright 1994). In particular, new social movement (NSM) theory contrasts "new" movements (particularly feminism, ecology, and peace) with the "old" (workers') movements. The NSM approach stresses that what distinguishes new social movements is that their actors struggle to create new social identities, to open up democratic spaces for autonomous social action in civil society, and to reinterpret norms and develop institutions (Cohen 1985; Scott 1990). Ideological issues are important in that the new movements are understood to function primarily as transmitters of ideas or what Melucci (1985) calls "new media" that carry a symbolic or cultural message. NSMs are interpreted as trying to bring about cultural change in civil society by transforming values, lifestyles, personal identities, and symbols (Melucci 1985).

The sustainable agriculture movement shares many characteristics with new social movements (Allen and Sachs 1993). This becomes particularly clear when examining the distinctive approach to experientially based knowledge that Hilary Wainwright (1994) identified in her analysis of several new social movements in Europe. In an analysis of the feminist movement, the student movement of the 1960s, radical trade unionists, and other recent movements, Wainwright (1994:5) argued that part of what makes these so-called new social movements distinctive (or new) is their assertion of alternative views of knowledge. An important element of these alternative views includes their critique of the dominant character and organization of what counts as valid knowledge and a legitimate source of authority. And as Wainwright argued, out of that critique grew innovation. In particular, participants in many new movements assert the validity of practical, experiential knowledge. While other analysts see such knowledge as that which we know but cannot tell, Wainwright argued that personal knowledge is transmissible and can be socialized. In the movements Wainwright studied, actors develop innovative ways to overcome the limits of their personal knowledge as they create and draw upon opportunities for gathering and exchanging practical as well as theoretical knowledge. In the process, they often bring to light knowledge that has otherwise been kept secret, ignored, or underused.

Such social cooperation is closely associated with participatory democracy. Wainwright (1994:11) argued that this democratic approach to knowledge and emphasis on "horizontal forms of organizing" are critical to achieving purposeful and effective social change in new movements. For Wainwright (1994:58), these insights have important ramifications because "if knowledge is a social product then it can be socially transformed through people taking action – cooperating, sharing, combining knowledge – to overcome the limits on the knowledge that they individually possess."

Like Wainwright, Ron Eyerman and Andrew Jamison (1991) argued that both "new" and "old" social movements are more than simple challenges to power; they are also creative engines that produce new knowledge and new organizational forms and principles. Eyerman and Jamison (1991:55) contended that social movements involve fundamental shifts in the consciousness of participants, a process in which theory and practice dynamically inform each other. They also see organizations as vehicles for transporting and disseminating the movement's message and for providing "a space for new kinds of ideas and relationships to develop" (Eyerman and Jamison 1991:60).

The contributions of Wainwright (1994) and Eyerman and Jamison (1991) are useful for interpreting questions that have not been addressed sufficiently in previous social movement theory in general and in social studies of sustainable agriculture in particular: Do issues raised by and within social movements lead to the formation of new thoughts, ideas, and practices – new knowledge? If so, in what kinds of social spaces do such activities take place, and what is the meaning of that new knowledge for participants? How can collective activity centered around the creation and exchange of knowledge lead to social change? This study uses these theoretical formulations about the role of knowledge in social movements to explore these questions in the context of farmers' networks as local expressions of the wider sustainable agriculture movement.

The Importance of Local Approaches to Social Change

The observations that I have made above – that many local-level organizations are now involved in the creation of an alternative knowledge system outside of dominant institutions; that local knowledge is a potentially powerful but too often neglected resource; and that social movements can involve social experiments into new ideas, practices, and organizational principles – all relate to the fourth factor that motivated this study. Specifically, in considering how society will move toward a sustainable agriculture, I have come to attribute a great deal of importance to the ways that people can and do participate in positive social change at the local level and in their everyday lives.

My views on local agency have been shaped in large part from my own experiences as an activist and community organizer in several social movement organizations during the last ten years. In working on a range of environmental and agricultural issues, I have encountered a variety of approaches to social change with respect to environmental problem solving. All approaches have advantages and disadvantages. It seems healthy – and perhaps essential – that a social movement incorporate different people participating in different ways. Based on my own experiences and reflection, I have made several observations regarding the fundamental question that drives this work: How can those of us committed to sustainable agriculture bring about long-lasting social change with respect to social and environmental problems?

First, the scale at which social problems are defined seems to be critical because the content of appropriate solutions is often implied by the definition of what problem needs to be solved. Karl Weick (1984) has shown that if the scale of a problem is defined in extremely broad, global terms, people often feel frustrated, helpless, and overwhelmed by the magnitude of the task. In thinking about today's agriculture, it is easy to be daunted by massive ecological and social problems and by the power of the national and international forces shaping the industrialization of agriculture.

Ultimately, however, sustainability is a local problem because sustainable agriculture must be applied and supported in particular places. There is a growing recognition that solutions to agricultural problems can come from the local level. As Peggy Barlett noted in her study of how the farm crisis of the mid-1980s affected farmers in Dodge County, Georgia, "it is important to avoid portraying farm families as victims or as helpless pawns; local cultures are capable of response and resistance to external forces" (1993:16). Similarly, Sarah Whatmore (1994) argued that theorists who explain the globalization of agriculture as an inevitable rather than a contested process effectively silence and deny social agency to rural areas that resist being incorporated into the international market economy. Clearly, by taking incremental steps in their everyday lives people can mobilize resources in their localities to achieve whatever is presently possible in the pursuit of long-term goals.

A second and related observation is that grassroots organizations are a critical component in social change efforts. In the early 1990s, I worked as a community organizer on agricultural and natural resource issues with the Northern Plains Resource Council (NPRC) in Montana. From the members and leaders of that twenty-five-year-old organization I learned that ordinary people can significantly affect the directions of their own communities, their state governments, and even the federal government by working together to achieve common goals. A democratically run group, guided in large part by the tradition of community organizing associated with Saul Alinsky (1971), NPRC takes an unabashed, often confrontational approach to social change. In community organizing it is important to build the power of the organization, to know the opponent, to devise effective strategies, and to do whatever it takes – within nonviolent and legal limits – to win. NPRC has won on a great many issues over the years and has developed the leadership capacities of its members. Community organizations do not have all the answers, but as Harry Boyte (1984) pointed out, they do suggest a process through which a communitarian and democratic spirit can be reinvigorated into political life in the United States.

Third, the community organizing approach seems to work particularly well when there is a clear target and a decision maker who can give the group what it demands. But it is sometimes difficult to identify and directly confront a specific decision maker for agricultural issues. For example, NPRC has taken on some of the most powerful, multinational corporations in the world by working in alliance with other organizations to push for antitrust actions against the meatpacking corporations that dominate the livestock market. In such a case, the focus is necessarily shifted from the local level to the national and global levels. At these levels the decision makers that can meet the demands of the organization are less identifiable, less accountable, and therefore much harder to influence effectively.

A final observation is that the regulatory approach to social change, the focus of many professionally oriented environmental groups during the last twenty-five years, has been a necessary, but perhaps insufficient, strategy. Working for the passage of legislation can be a rewarding process that stimulates public discussion about the impacts of conventional agriculture. Legislative strategies can also involve citizens in problem solving as they testify in front of committees and pressure their legislators to support (or oppose) a particular bill. But even passage of the best legislation does not ensure that the responsible governmental agency will have sufficient resources or the political will to enforce the law adequately. Enforcing environmental laws in the agricultural arena is particularly problematic because there are so many land users to monitor for compliance and because noncompliance is difficult to prove because pollution is often found far from the original source. Inadequate enforcement also means that all too often environmental organizations must resort to litigation either against a company that is violating a law or against a governmental agency that is failing to implement or enforce a law for which it is responsible. When successful, litigation can be a powerful tool. But litigation is expensive and tends to involve directly only a fairly narrow group of people in the social change effort, primarily professional activists, lawyers, and scientists who act as expert witnesses. During the last.two decades, environmental groups have accomplished a great deal through passage of legislation, participation in the regulatory processes of governmental agencies, and litigation. As the contemporary erosion of environmental law illustrates all too boldly, however, these achievements may not withstand the test of time.

These thoughts about social change were on my mind when I arrived in Wisconsin in 1992 and started to learn about the sustainable farming networks. I quickly began to suspect that the networks might represent an approach to social change that was distinctly different from my previous experiences in social movement organizations. Specifically, I began to ask whether lessons might be learned from the creativity of farmers who have survived outside of the conventional systems and who are actively engaged in practicing alternative production and marketing techniques. And I wondered if a key element to producing social change in agriculture is the exchange of knowledge among those farmers in their local networks.

The profound need for radical change in agriculture is now being voiced loud and clear, and visions of a sustainable future are beginning to be articulated. Less clear is the road that could move us from here to there. Addressing the knowledge questions in sustainable agriculture is undoubtedly a critical component of the effort. With the rise of the institutions of agricultural science, farmers came to be viewed as the recipients of knowledge generated by professionals we call scientists. But the social movements of the 1960s and 1970s reinvigorated critiques of dominant agricultural science and the industrialized, globalized agriculture it has made possible. Efforts to reorient agricultural science toward lines of inquiry and research methods more favorable to sustainable agriculture are being complemented by the creation of an alternative knowledge system that largely functions outside of the formal institutions of agricultural research. Yet little is known about how this alternative knowledge is developed and shared at the local level. This study was an attempt to learn more about those alternative sources of creation and exchange of knowledge.

THREE

Exploring the Landscape of Sustainable Alternatives in Wisconsin

Soon after moving to Wisconsin in the fall of 1992, I went with two companions from the University of Wisconsin – Madison to a Corn Plot Tour sponsored by the Southwest Wisconsin Farmers' Research Network (SWFRN). The SWFRN, the first sustainable farming network in Wisconsin, was in its sixth year. The purpose of the tour was to describe an on-farm research trial comparing the performance of six hybrid corn seed varieties. About twenty people – mostly farmers as well as a few rural advocates and a seed salesman – gathered beside the cornfield to hear about and discuss the results of the variety test. The discussion covered a range of factors that the farmers felt were important to consider in evaluating the corn varieties. For example, the corn variety had to work on organic dairy farms that used on-farm nutrients (legumes and manure) and mechanical weed control (rotary hoeing and cultivation). They were particularly interested in the relationship between corn yields and costs of production, and the network coordinator handed out a summary table comparing the corn varieties with respect to these factors. The group also considered less quantifiable factors. The host farmer, for instance, was interested in how easy it would be to harvest the corn using his thirty-year-old combine, which had its own quirks. And in hopes of reducing the need for mechanical weed control, the farmers judged how much shade the corn variety provides because shade minimizes the amount of light that can get to weeds and thus limits their growth. The farmers seemed to move easily among topics – from genetics, to economics, to mechanics.

At the end of the tour, the host farmer asked me and my companions if we wanted to look at what he thought was really much more "exciting" than the corn trials. We followed him out of the cornfield to the lush pastures that this farm family had begun to manage intensively just a few years before. The rotationally grazed pasture had been subdivided into small paddocks laid out around the contour of a gently sloping hill shaped like a bowl. The farmer pointed out the different species growing in the pastures, and he conveyed an infectious excitement as he shared with us what he felt were the benefits of rotational grazing over corn silage for feeding dairy cows, even if the corn is grown organically. After our short and lively conversation, he opened the gate where the cows were grazing, and they followed him down the lane and back to the barn for the evening milking. We returned to Madison.

That day stands out in my mind because it inspired me to learn more about the social activity around sustainable farming networks in Wisconsin. I began by becoming acquainted with the individuals and organizations working on sustainable agricultural initiatives. I started this process in earnest in 1993. Yet this story of farmer networking began in the mid-1980s, when a small group of farmers collaborated with rural activists to form the SWFRN. Since the formation of that network, farmers and other rural advocates have encouraged and facilitated the creation of many networks around the state. In this chapter, I describe the general development and growth of sustainable farmer networking in Wisconsin. By sustainable agriculture "network" I refer to those groups formed primarily for the purpose of organizing opportunities for farmer-to-farmer communication about practices and ideas associated with sustainability.

After preliminary research, I focused my inquiries more specifically on two farmer networks that were in their formative stages when I began to attend their events in early 1993: a network of dairy farmers practicing rotational grazing and a network of women interested in sustainable agriculture. For each of these two networks, I shall discuss the process of network formation and the reasons why the group leaders initially felt the networks were needed. Woven into this story is an account of how I selected these two particular networks for further study and how I approached gathering, analyzing, and interpreting the information that is the basis for this narrative. In these ways, I try to set the stage for exploring more fully in subsequent chapters how these sustainable farmer networks function, what they mean to those involved in them, and the implications of these networks to those engaged in similar activities elsewhere.

The Emergence of Networking in Wisconsin

During the winter of 1986, four farmers from southwestern Wisconsin traveled together to a conference in LaCrosse, a town located on the bluffs above the Mississippi River, which forms the border between Wisconsin and Minnesota. The two-day meeting was sponsored by the Rodale Institute, the Pennsylvania-based organization that has long played a pivotal role in promoting organic and other alternative forms of agriculture. For the four farmers, the LaCrosse meeting was significant not so much for what happened there as for what it inspired them to do afterward. The meeting sowed the seed for what eventually became known as the Southwest Wisconsin Farmers' Research Network, the history of which is worth briefly recounting here for two reasons. First, the SWFRN was a catalyst for the current networking system in Wisconsin. Second, the motivations behind and the activities of that first network provide a useful reference point for discussing how the Wisconsin sustainable farming networks have remained the same and how they have changed over time. In describing the origin of farmer networking in the sustainable agriculture movement in Wisconsin, I rely principally on the work of Margaret Krome (1988), an activist involved with the group, as well as on what I learned about events from some of the people involved.

While driving back to their homes from LaCrosse, the four farmers reflected on their experiences at the meeting. They had enjoyed the opportunity to talk with others interested in farming with fewer pesticides and synthetic fertilizers. And they had learned about the experimental field trials that the Rodale Institute was conducting on its demonstration farm near Kutztown, Pennsylvania. Their experiences triggered a discussion about the lack of reliable information on alternatives to conventional farming techniques. These farmers felt that the land-grant university system had too rarely studied questions of concern to those who wanted to farm without agricultural chemicals. In addition, they were troubled that research conducted under controlled experimental conditions did not adequately reflect the variable and complex situations that farmers confront every day on their farms. Compounding the problem – from the viewpoint of these farmers – was that university extension agents often seemed unwilling or unable to help farmers obtain whatever information on alternatives might be available.

More important than the critical discussions this small group had regarding the agricultural research system was that these farmers were determined to do something about the situation. Inspired by the Rodale Institute's research trials, the four farmers and another who joined them later decided to conduct experiments on their own farms. They wanted to answer questions that they identified and that they felt had been ignored by the established institutional mechanisms for generating and disseminating information to the agricultural community.

To act on their idea, these farmers secured the help of the Wisconsin Rural

Development Center, a nonprofit advocacy organization formed in 1983 by farmers, rural activists, and clergy concerned about the impact of the farm crisis on Wisconsin's land and rural people. This organization was a logical one to turn to; some of these farmers had been board members of the WRDC and, at that time, one of them also worked for the WRDC. The WRDC was able to secure funding from a private foundation to finance the on-farm trials and to cover staff time in coordinating activities of the farmer network (in addition to supporting other sustainable agriculture efforts the WRDC was involved with at the time).

Although the WRDC mobilized resources that would not otherwise have been available to help the farmers achieve their goals, the involvement of that organization apparently did not usurp the active role of the farmers themselves in producing and sharing knowledge. Rather, a core principle that guided the farmer network and the WRDC was that farmers and the specific on-farm conditions under which they operate should be central to developing research projects on agricultural production, whether those projects are done by a farmer network or a land-grant university. According to Krome (1988:2), who was on the WRDC staff, the network members felt that farmers "know best what information will be most useful to them, and they have a wealth of information that can enrich a research effort and make it more efficient." The project's organizers also thought that farmers should be active in communicating the intentions, methods, and results of the research to a farm community that was often skeptical of alternative ideas. This commitment to developing farmers' capacities to produce and disseminate new knowledge about sustainable agriculture still lies at the heart of farmer networks.

In recognizing their own capacities as producers of knowledge and in articulating their critique of the agricultural research system, the SWFRN farmers did not turn away from science per se. Instead, the network members actively sought collaboration with people in established research institutions, principally at the University of Wisconsin. The farmers welcomed advice from what they considered to be a few "unusually interested" (Krome 1988:8) extension agents who were willing to help. Krome's (1988:8–9) account of the first two years of SWFRN indicates that the network emphasized trying to establish cooperative relationships with individuals in the university system and to generate "methodologically valid research" knowledge that would convince the academic community of the viability of certain alternative practices.

With the help of the WRDC and several sympathetic extension agents, the network farmers decided that their first project would be to conduct on-farm trials designed to demonstrate the economic benefits, that is, cost savings, of growing corn with reduced applications of purchased fertilizer and to evaluate the contributions of nitrogen to corn yields from plowed-down alfalfa and cow manure. During the first year, 1986, the network set up trials on five different farms. The trials demonstrated that the yields of corn whose nitrogen came solely from plowed-down alfalfa and manure did not differ significantly from yields on land where commercial nitrogen had been applied in addition to these other, nonpurchased sources. Thus the network found that reducing or eliminating the use of commercial fertilizer lowered the cost of producing corn without lowering yields, and it minimized the application of fertilizers that might cause groundwater contamination, a serious environmental problem associated with agriculture in southwestern Wisconsin. In 1987, six farmers repeated the same research trials on each of their farms and two other farmers conducted similar trials related to phosphorous and potassium applications.

Meanwhile, two other network members demonstrated the use of intensive rotational grazing on their dairy farms, although they did not conduct formal or systematic experiments. They were among the first in the state to practice this technique, which is in many ways a major departure from conventional dairying in Wisconsin. One of the two SWFRN grazers had learned of the technique from a family member who was practicing it on a Vermont dairy farm. Since these initial SWFRN demonstrations, many more dairy farmers have switched to this method of feeding their dairy cattle. The SWFRN members played an important leadership role in introducing, demonstrating, and promoting this now increasingly popular technique.

The SWFRN used several mechanisms to share its findings with other farmers. It held sustainable agriculture field days each year so that participating farmers could present and discuss the results of their study with other farmers and interested parties from around the state. These events were popular, especially during the network's early years, and the SWFRN received a considerable amount of media attention. To promote the network's activities, members became adept at working with and identifying issues of interest to the press. For instance, one family farm in the network received much attention when it won a regional corn profitability contest in which it was the only participant that did not use pesticides or commercial fertilizers. The network attributed much of its success to the interest of the state's agricultural press.

During the network's first year, the SWFRN farmers and the WRDC realized the benefits of farmers working together to identify and solve problems collectively. They also recognized the need to establish institutional mechanisms for allocating resources toward farmer-based research in alternative farming systems. As a result, the WRDC successfully coordinated a grass-roots campaign seeking legislative approval and funding of a Sustainable Agriculture Program to be housed in the Wisconsin Department of Agriculture, Trade, and Consumer Protection (WDATCP). Members of the SWFRN were persuasive advocates for the program in the Wisconsin state legislature because they could refer to their own on-farm research as a successful demonstration of how to reduce use of energy-intensive fertilizers and pesticides.

One of the first of its kind nationwide, the Sustainable Agriculture Program in Wisconsin was established late in 1986 for the purpose of allocating grant funds on a competitive basis to individuals or groups who proposed networking, research, or demonstration projects that would serve "to improve simultaneously the environmental soundness and the profitability of Wisconsin's farming and agriculture, to increase the energy self-sufficiency of the Wisconsin agricultural system, and to empower Wisconsin farmers to contribute directly to both" (WDATCP 1991b:2). The competitive grants program was supported with money received from a federal court settlement that required compensation to Wisconsin citizens for petroleum product overcharges made by oil companies during the 1970s and early 1980s. The legislature designated $2 million to be allocated to this sustainable agriculture demonstration program.

In sum, the SWFRN and the WRDC played a strong and deliberate role in laying a foundation for the further development of sustainable farmer networking in Wisconsin. They did this both by providing a visible example of the potential of farmers themselves to generate knowledge for sustainability and by achieving their policy goal of getting the Sustainable Agriculture Program established. As Krome (1988:9) observed:

When the [Southwest Wisconsin Farmers Research] Network began two years ago, it had to define the arena of sustainable agriculture in Wisconsin, awaken interest in it, and give it legitimacy in the eyes of both farmers and policy makers. Today, the movement has taken on a life of its own. Network contributions were a healthy mixture of practical research questions and policy strategy. Each one fed the other, the on-farm research giving meaning to the policy issues, and the policy successes giving birth to a live and growing movement. The farmers and other members of the Network look on their contributions with justifiable pride.

Indeed, the growth and diversity of farmer networks in Wisconsin today is a continuing testament to initiatives in the mid-1980s.

Networks Take Root

The role of the Sustainable Agriculture Program as a catalyst for developing networks seems to have been strongest during its first five years. In its fifth year of operation, 1992, the Sustainable Agriculture Program awarded a total of $389,854 to twenty-four grant recipients. Eight of these recipients were farmer networks receiving grants ranging from $14,440 to $31,051 and totaling $177,196, or 45 percent, of the program's funding. During subsequent years, funding for networks decreased. By 1995 the sustainable agriculture advisory council recommended funding for fifteen projects totaling over $225,000. But only two of those projects were to establish networks, and their awards totaled only $20,500, or 9 percent, of the money allocated. Although the reasons for the decrease in funding to networks is unclear, program staff suggested that the advisory committee was concerned that too much funding had been used to cover salaries of network coordinators. Changes in priorities and budget cuts also meant that less money became available for networks over time. The oil overcharge funds that funded the program ran out in 1997. As a result, Governor Tommy Thompson eliminated the program's staffing positions in his 1995–97 budget, and the Sustainable Agriculture Program dispensed its final $100,000 during the 1996 funding cycle to cope with the funding losses (Fyksen 1994b; McNair 1995).

Surprisingly, despite the decline in state support for sustainable agriculture networks over the years, the number of these groups continued to increase. Ten networks were established between 1986 and 1990 (WDATCP 1991a), but another twenty were established between 1991 and 1995, indicating their growing popularity and their ability to continue with little or no direct funding from the state. Appendix A lists the thirty sustainable farmer networks that I identified as of 1995.

The primary purpose of networks is to organize opportunities for generation and exchange of information among farmers. These informal groups tend to be structured relatively loosely and simply, and they use a variety of organizational approaches to support the exchange of information and ideas related to sustainable agriculture. For instance, some networks cover only one or two counties, while others encompass a larger territory. For the most part, networks seem to have moved away from conducting the systematic on-farm research done by the SWFRN in its early years. Instead, network activities often include one or more of the following: hosting field days or farm tours, organizing winter workshops or informational conferences, and publishing newsletters and bulletins.

Networks differ in whether they are organized around a diversity of sustainable agricultural initiatives or around a fairly specific topic. The first approach has been adopted by networks in which members find common ground even though their enterprises differ in the products produced, the farming practices adopted, or the marketing channels used. What members do share in these diverse networks is a commitment to exchange of knowledge about sustainable agriculture among farmers. Accordingly, events cover a wide variety of topics.

An example of this first approach is the Western Wisconsin Sustainable Farming Network (now known simply as the SFN). In 1993, the SFN held a series of workshops on sustainable agriculture. The twice-monthly winter sessions covered a range of topics that illustrate the mix of practices and issues addressed by the group: "Soil Health and Productivity," "Sustainable Weed Control," "Preparing for the 1995 Farm Bill," "Financial Fundamentals on the Farm," "Developing and Sustaining Your Health on the Farm," "Rotational Grazing: Techniques, Technologies, and Problems," "Affordable Ways to Start Farming," "Issues and Problems on the Sustainable Horizon," "The Urban Food Consumer," and "The Place of Sustainable Agriculture in the Economy." In a direct challenge to the authority of agricultural science, the network chose to call these events "winter institutes," reminiscent of a term used to describe similar events organized by farmers as an alternative to the land-grant colleges when they were first established in the late nineteenth century. The coordinator of SFN described the overriding purpose of these winter institutes:

[I]t has become clear that "college knowledge," the on-the-face-of-it foolish idea that middle-class, professional people comfortably removed from the day-to-day practices of farming could know more, and better, about farming than farmers themselves is a bankrupt (and bankrupting) idea. Farmers in the 1870s knew it, and took over the farmer institutes accordingly. Sustainable farmers in the 1990s know it. Hence our Institute, which is simply a revival of the notion that farmers are the best source of information about farming. (Schaefer 1993:2)

A second and more common type of network focuses around a specific topic. For example, two networks – one in the area around Madison, Wisconsin, and the other in the Wisconsin-Minnesota border region near Minneapolis–St.Paul–concentrate on educating people about community-supported agriculture (CSA). An increasingly popular way of organizingorganic farms, CSA is a mechanism that allows producers and consumers to forge new relationships with one another around their local food system. On a CSA farm, consumers purchase shares of a farm's harvest, farmers receive money to put in the season's crops, and the farm's produce is delivered to its shareholders on a weekly basis throughout the growing season. In the CSA networks, farmers share knowledge about developing and maintaining successful community farms even though their farms often compete for shareholders (Ostrom 1997).

The two networks that are the focus of this study constitute examples of each of these two major types of networks. The Wisconsin Women's Sustainable Farming Network is an example of a network that incorporates a range of topics in its activities and is made up of members who are engaged in a diversity of sustainable agricultural initiatives. The Ocooch Grazers Network is an example of a network organized around a specific topic, in this case intensive rotational grazing and its associated practices. Such grazing networks have become the most prevalent type of network in recent years because of the increasing popularity of this technique among dairy farmers. As of spring 1995, eighteen grazing networks were dispersed across Wisconsin's agricultural landscape.

Networks in Wisconsin are connected to a wider community of organizations and individuals involved in sustainable agriculture. Some annual conferences act like loose federations of network members, other alternative farmers, interested researchers, and rural advocates. Most important among these conferences are the Wisconsin Grazing Conference, organized by Grass Works, Inc., and the Upper Midwest Organic Farming Conference, organized by the Wisconsin chapter of the Organic Crop Improvement Association and other organic farming groups. There are also other, more formal organizational participants in the sustainable agriculture movement in Wisconsin. For example, there are organizations that conduct public education and advocacy in the policy arena in order to pursue one or more movement goals (e.g., the Churches' Center for Land and People, Michael Fields Agricultural Institute, Wisconsin Farmland Conservancy, Wisconsin Rural Development Center); specialize in certifying organic farms (e.g., the Wisconsin chapter of Organic Crop Improvement Association); and collectively market organic products (e.g., the Coulee Region Organic Produce Pool). I found that many network farmers also participate in some of these more formal organizations or the events that they sponsor.

Origins of Two Wisconsin Networks
The preliminary phase of this research involved several steps to become better acquainted with the individuals and organizations working on sustainable agricultural initiatives in Wisconsin. For example, I read regularly the two prominent agricultural newspapers published weekly in Wisconsin,

Country Today and *Agri-view*. From those papers, I collected articles covering a wide variety of topics related to sustainable agriculture and the people involved in it in the state. I also identified the various organizations working on issues related to sustainable agriculture and joined the mailing lists of those groups to keep abreast of their activities.

Whenever possible, I attended statewide conferences and locally oriented events sponsored by networks and other organizations. Gaining entry to these events was not a problem because they were all open to the public. I introduced myself as a researcher interested in the role of networks in the sustainable agriculture movement in Wisconsin. I was forthright about my own commitment to creating a more sustainable agricultural system, and I think this helped me establish rapport with many of the people I met. Drawing on my previous experience as a community organizer, I felt comfortable pursuing contacts and developing relationships with the farmers and activists I met. Many of these initial contacts proved to be valuable throughout the research project.

Preliminary observations and informal conversations with people in Wisconsin's sustainable agriculture movement led me to the two networks that became the focus of this study – one network for dairy farmers practicing intensive rotational grazing, the other for women farmers practicing a variety of techniques associated with sustainable agriculture. Along with specific input from several activists and farmers, this preliminary work was useful in helping to identify and develop specific research questions. What follows is a brief description of the origins of each network, their primary activities, and the reasons why these two research sites were chosen for further study. (A detailed organizational profile of each network is provided in subsequent chapters.)

The Ocooch Grazers Network

In the winter of 1993, a group of dairy farmers gathered at a meeting room of the public library in a small town located in the heart of the hilly region of southwestern Wisconsin. Mike Cannell and Jim Brown had called the meeting to discuss the possibility of forming a local farmer network for those interested in intensive rotational grazing. Having been among the first in the area to practice the technique, Mike and Jim had been called upon individually many times to help others get started with grazing on their own farms. According to Mike, he and Jim "knew there was a need for [a network] because both of us were spending too much time on the telephone trying to individually serve people. . . . And so we said, well, if we can get a grazing network going, that will allow a lot of people to learn all at the same time. . . . And people will learn that they shouldn't be calling us individually, they should be going to the pasture walk and asking questions and then everybody can gain by the answer. So we realized that we could create an educational vehicle that would work."

The network coordinators were also confident that they had the ability to organize the group. In part, they looked to farmer discussion groups in New Zealand as a model. While the U.S. dairy sector pursued the development of confinement feeding, dairy farmers in some other parts of the world adopted and further developed pasture management principles elaborated in the 1950s by the French farmer and scientist André Voisin [(1959) 1988]. New Zealand is the most notable example of an agricultural system relying almost exclusively on permanent pastures, and it has one of the lowest-cost dairy industries in the world (Murphy 1991). Not only do New Zealand dairy farmers rely on permanent pastures to feed their livestock, they also rely on one another to improve their grazing practices. As Jim explained:

I get a magazine from New Zealand called *Dairy Exporter*, you know a monthly magazine, and it said certain discussion groups [in New Zealand] celebrated their thirtieth anniversary. And I got to reading it, and it said they have a meeting on each member's farm once a month, and it's not a meeting to cut people down, it's a meeting to build people up. So I made a copy of that article, and I mailed it to Mike. . . . I said, "Think about this." And when he got that copy in the mail, he called me up and said, "That's it, we're starting a group."

Mike was familiar with farmer networking from his own involvement in organizations around Wisconsin, including the SWFRN. Describing the confidence these experiences engendered, Mike said:

I knew we could do it. I had been on the steering committee that developed the Wisconsin Rural Development Center, and I was one of the four farmers that created the Wisconsin Family Farm Defense Fund. We incorporated it. And, of course, I knew my father had started the [local chapter of] Farmers' Union. And so I knew that I had the ability to do this. I owned a computer, and I owned a copy machine, and I had enough knowledge about . . . written communication. And I knew that it's much more efficient if you reach out and touch people a little bit more frequently . . . than if you touch them once a year and give them so much they won't read it anyway. . . . I knew that if I would attach a little bit of education to the back of each pasture walk [reminder], it would trigger their curiosity and we'd kind of carry them along, see, and then build on this knowledge base. So I knew that that would work.

Intensive rotational grazing differs markedly from the dairy systems that are the convention in Wisconsin and other regions of the United States. As a result of a hundred-year process of scientific investigation and technological invention, conventional dairying focuses on maximizing milk production by feeding stored forage to cows that are confined largely to the barn and its immediate environs (Fales 1994; Murphy and Kunkel 1993). On such confinement-based farms, dairy cattle are fed year-round by the farm operator, and most land is typically used for producing crops such as corn, alfalfa, and soybeans. Major capital outlays are required for the fuel, equipment, silos, and other inputs needed to produce and store feed, to move the feed to the cows, and to remove and redistribute manure (Murphy and Kunkel 1993). These costs, coupled with decreased milk prices, have greatly reduced the profitability of confinement dairy systems, particularly for many family-owned and operated dairy farms with herds of fewer than one hundred cows (Rust et al. 1995).

In contrast, the new "grass radicals," as journalist Joel McNair (1992b) has called them, are asking how they can manage land as permanent pastures so that animals can harvest high-quality forage for themselves during as much of the year as possible. Unlike traditional pasture management systems for dairying that use continuous grazing, these grass radicals have adopted the grazing management principles originally developed by Voisin. Terms such as *management intensive grazing*, *intensive rotational grazing*, and *Voisin-controlled grazing management* are used to describe management approaches that employ very similar principles; however, Wisconsin farmers practicing this technique most commonly use the term *rotational grazing*.

Rotational grazing involves two key elements that Voisin ([1959] 1988) advocated to ensure that pasture plants have a chance to photosynthesize and replenish energy reserves after each grazing. First, the farmer divides land into small areas, or paddocks, using portable fencing and watering systems and then rotates animals through these paddocks. Second, employing short grazing periods with high stocking density on these paddocks ensures that nutritious and palatable forage is available for the animals and that pasture plants have sufficient time to recover and are not overgrazed. Voisin argued that such attention to time was the critical factor distinguishing his management principles from other grazing strategies.

Historically, pastures in the United States have been poorly managed. When dairy production for larger and more distant markets accelerated in the first half of this century, there was a need to improve milk yields and extend the lactation time of the dairy cow. To meet the rising demand, forage from pasture was rapidly replaced by the intensified production of storable feed crops such as corn silage (Fales 1994). Work of agronomists and dairy scientists in the land-grant universities facilitated and hastened this process. Some agricultural scientists experimented with varieties of grazing man-

Unlike conventional dairying, where the farmer brings feed to cows kept in confined areas, cows in a rotational grazing system go to the pasture to harvest forage for themselves during as much of the year as possible.

Rotational grazers subdivide pastures into paddocks by installing lightweight fiberglass fence posts and stringing these posts with polywire. At the corner of the paddock, shown here, a grazer demonstrates connecting the wire to a more permanent fence line that will provide the necessary electric charge.

agement, including Voisin's, throughout the 1950s, but they failed to give adequate attention to the amount of time animals spend in a paddock. As a result, trials conducted by H. J. Larsen and R. F. Johannes (1965) at the University of Wisconsin's Marshfield Experiment Station in the late 1950s and early 1960s led agricultural scientists to believe that it was best to feed cows on stored forage even during the summer. McNair (1993:2) and others have maintained that the results of that study and the "attending publicity played a major role in the evolution of the state's dairy industry away from pasture and toward confinement feeding."

Since the 1950s, agricultural scientists in the United States have pursued research questions almost exclusively related to confinement systems. As a result, the formal institutions of agricultural research and extension had little to offer dairy farmers interested in the unconventional technique of rotational grazing. As a dairy specialist from the University of Wisconsin told grazers at a conference, dairy scientists "don't understand the basics" associated with grass-based dairying. So when more and more farmers began to make a transition from confinement to pasture-based feeding during the late 1980s, the new grass radicals sought information and support mainly from the networks that many of these same farmers had helped to develop earlier, such as the SWFRN.

At that first meeting called by Mike and Jim, members of the new group made decisions about how they would try to help one another build on what Mike referred to as a "knowledge base." The group scheduled a series of pasture walks to be held at a different farm each month during the coming grass season, roughly April to November. These monthly pasture walks, which typically last for two or three hours, became the primary mechanism that this grazing network used to facilitate exchange of knowledge among farmers. Between grass seasons, the network held one or two winter meetings at which they scheduled walks for the following year, listened to invited speakers, and shared a potluck meal. This organizational structure is fairly typical of the grazing networks that have been formed around Wisconsin in recent years.

When Mike and Jim first envisioned the network, they imagined bringing together about fifteen to twenty dairy farmers interested in grazing from two neighboring counties, Vernon and Richland. But as it turned out, many farmers were "hungry for information" according to Jim, and a fairly large crowd showed up at the first meeting. In only two months' time, the mailing list included 102 people and has remained at a comparable size ever since. Of course, not everyone on the mailing list comes to network events. One walk I went to drew as many as a hundred people, but it was held at a farm well known for its innovative practices and success with grazing, and it was cosponsored with several other networks. More typically, there were about twenty or thirty people at the pasture walks I attended. Of the twenty core members who came regularly, five were women who usually attended with their male partners (although one woman occasionally came by herself).

The final major agenda item on that winter day in 1993 was to choose a name for the group. Mike and Jim suggested – and the others who were gathered there agreed – that the new group be called the Ocooch Grazers Network. "Ocooch" is a Native American word for the mountains of southwestern Wisconsin. The geography of the area is defined more by what did not happen there in earlier geologic epochs than by what did occur. While the rest of what is now the state of Wisconsin was dramatically altered by the repeated advance and melting away of glaciers beginning some million years ago, the higher elevations of southwestern Wisconsin were surrounded but never touched by the glaciers that flowed more easily elsewhere (Vogeler 1986). Today, this area is known as the Driftless Area because no drift – glacially deposited sand, rocks, and boulders – was set down. The Driftless Area, with its steep hills and deeply cut valleys, contains some of the most rugged and potentially erodible terrain in the state. The word *grazers* in the network's name refers to the farmers who practice rotational grazing. Some people use the word *grazers* to refer to livestock that graze pasture, and they reserve the term *graziers* for the farmers who manage the livestock. In this book I follow the spelling and usage adopted by this particular network. Thus the term *grazers* refers only to farmers who practice rotational grazing.

The Wisconsin Women's Sustainable Farming Network

Women have always been productive participants on farms. Nevertheless, men tend to control land, labor, and capital in the U.S. food and agriculture system, including on the vast majority of family farms (Sachs 1983). The proportion of farmers in the United States who are female appears to be increasing. One measure of this increase is the U.S. Census of Agriculture, which allows for identification of one operator per farm, that is, the person who has primary responsibility for running the farm. Until 1978, the gender of the operator was not recorded, and thus the census provides little historical perspective on women's involvement as primary operators. According to recent census figures, however, in 1978 the number of female farm operators was 128,170 and the number of male operators was 2,350,472. By the 1992 census, the number of women farming had increased to 145,156 while the number of men farming had fallen to 1,780,144. Thus over the fourteen-year period, the proportion of women farmers having primary responsibility for a farm grew from 5.2 to 7.5 percent nationally (U.S. Bureau of the Census 1995).

Similar figures are available for Wisconsin. According to the 1982 census figures, there were 3,256 female farm operators in Wisconsin and 78,943 male farm operators. By the 1992 census, the number of female operators had increased to 3,823 and that of male operators had fallen to 64,136 (U.S. Bureau of the Census 1995). During that ten-year period, the proportion of women having primary farm responsibility grew from 3.9 to 5.6 percent. Some members of the Wisconsin Women's Sustainable Farming Network (henceforth referred to as the Women's Network) are among this group of women having primary responsibility for a farm. Other members of the network shared responsibility with a partner and may or may not have considered themselves to be the primary operator of the farm. Most important, all network members either identified themselves as or aspired to be farmers despite the common association of the term *farmer* with men.

The idea for the Women's Network was born out of the activities of another network, the Western Wisconsin Sustainable Farming Network. By the time Diane Kaufmann saw the newspaper advertisement for the first organizational meeting of the SFN in the spring of 1991, she was well on her way to realizing her lifelong dream of becoming a farmer. Diane's desire to farm felt natural to her when she was growing up not very far from where she lives today. Yet she did not feel that farming was a viable option to her because of her gender. Nonetheless, farming remained attractive to her:

My earliest memories of being on my grandmother's farm, even a lot of the smells, the smell of the hay mow and the cows, it just all seemed so pleasant to me. Even the cows. Being in the outdoors there, her big gardens, and she was an organic gardener way back then. All her flowers, the big pine trees around the house, birch trees on the hill, it was just kind of that idyllic Wisconsin farmstead, with the creek running through the farm where we could go down and play. . . .

Those are all childhood memories, Mom taking us down to the creek to go swimming. But I really liked the flow of the work too. My uncle let us be gofers for him, and I helped hay and stuff like that. Maybe because I didn't have to do it day in and day out at that point in time, it just seemed like it was fun and not work. I just always identified with that. It just seemed like a very natural thing to do. In like fourth grade, I had this black fuzzy sweatshirt, and I'd go down in the barn with it. . . . It would just gather up all that smell. And then, of course, we would stop at my other aunt and uncle's home in town on the way home, and they would just kind of pick me up and put me outside. But to me it was just, "Ah the farm."

Then as I grew older, even in high school, what I really truly wanted to do was to go to [the University of Wisconsin at] River Falls in the Dairy Science program or something like that and be a farmer. But that was 1969, and women didn't do that, especially if you didn't come from the farm. It just didn't seem like that was an option, and I didn't even mention it to anyone because I just knew I'd get laughed right out of the room. I wasn't strong enough I guess at that point to say, "I don't care; this is what I want to do." But yeah, actually what were my options doing that as far as a career? You know, go there [to the agriculture college] so you could find a farmer and marry him? That wasn't my goal. . . . So it's just been a dream I nurtured all through going other ways [in life] . . . just always wanting to get back to this area to be able to farm in some manner.

In the early 1980s, Diane did get back to the area of Wisconsin near where she grew up, and she convinced her husband and two children to buy a twenty-two-acre farm. By the time the first meeting of the SFN was held in 1991, Diane had been raising about eight hundred chickens on pasture in portable pens each summer for three years and had no problem selling them directly to people in her area. She also had been building her sheep flock for eight years, and she managed her sheep and poultry together in a rotational grazing system. That was also the first year she began to milk some of her ewes and to create one of the few sheep dairies in the United States. Many of Diane's farming practices are associated with sustainable agriculture so when she saw the meeting about forming a sustainable agriculture network announced in the paper, she decided to check it out.

The meeting was organized by a network coordinator who was then on the staff of the WRDC, which had received one of the grants from the Sustainable Agriculture Program to establish a network in the area. He asked people to indicate whether they were willing to be on the steering committee. As Diane watched other farmers volunteer, she noticed that they were all men. Then she volunteered herself.

During its first year, the SFN sponsored summer field days and winter institute sessions. According to its membership brochure: "The network's basic assumption was that farmers are the experts. Farmers are always interested in hearing from their peers about what works on their farms." While Diane participated in the SFN events, she noticed the lack of active participation by other women farmers. This observation began to raise questions for her:

I knew what I was getting out of the [SFN] was extremely valuable. And I knew that there were good people in that network, and they were helping me. So part of it was feeling that why aren't there more women here? . . . Are meetings like this intimidating to women? Are they just things you send your husband to since he's going to go or whatever? Or if you're not married, you're single or whatever, are you comfortable going there?

Diane decided to try to find some answers.

As part of a conference organized by the SFN in the fall of 1992 and with the support of the rest of the SFN steering committee, Diane organized a workshop aimed solely at women farmers. The workshop included several presentations. One highlight that women in the network commented on later was a presentation by Allan Nation, editor of the *Stockman Grass Farmer,* a monthly newspaper based in Jackson, Mississippi, and popular with grazers. Diane invited Denise Molloy to speak. Denise was slowly building a flock of sheep, marketing the high-quality wool to spinners, and selling lamb directly to local consumers. She also worked part-time as an occupational therapist. She told me:

Diane asked if I would speak on back care because she's had some back problems. . . . She thought other people would be interested, so I did a presentation on that. . . . Well, anyway, a lot of people came to that. It was really successful. And so then, the one comment that came out of it was that women wished, they wanted to get to know each other more, and there wasn't any time for networking at all. It was all speakers the whole day, and so we got a whole big list of names and stuff. And so then we had another one the following spring.

Diane felt that participants in that first workshop recognized that "there are things we know we can teach each other and learn from each other." She also recalled: "It was a totally different atmosphere. The women felt a lot of support and wanted it to continue." And continue it did. In the spring of 1993, Diane and Denise organized a day-long conference to which Diane invited me and which was the first one I attended. Like the many other Women's Network events I have observed since then, that spring day was filled with women sharing their experiences as women in farming, as well as exchanging technical information about their production and marketing practices.

From its inception, members of the Women's Network self-consciously experimented with their organizational structure and the activities they wanted to pursue. For example, to encourage the involvement of more women, they decided very early on to have a steering committee rather than one or two coordinators as is typical in other networks. The original core membership lived relatively close to one another in the Chippewa Valley area near Eau Claire, where many of the meetings took place during the period I studied. Two conferences a year held in that area also drew a few women from as far as five hours away, and the 1994 network membership list included sixty-four women from across the state. Membership numbers continued to grow, and the network divided into a northern and a southern chapter in 1995 to encourage more local activities. Responsibility for organizing an annual conference for all members now alternates between the two chapters each year. Such one- to three-day conferences, drawing anywhere from fifteen to forty people, have been the cornerstone activity for this network. The network also has a newsletter to which members contribute updates on their personal endeavors, and it has sponsored several field days and a few work parties.

Like the grazing network, members of the Women's Network wanted to provide opportunities for sharing practical information about alternative agriculture. Although the technical aspects of the grazers' network are relatively easy to typify, the agricultural interests of the women were extremely diverse because they pursued a wide range of endeavors. For example, during the period of this study, some women were raising sheep, and they were processing, spinning, and dyeing the wool to market a variety of products. One woman was managing about seven acres of organic basil, working with two partners to process and market frozen pesto around the Midwest. Another farmer was growing flowers organically, arranging and packaging bouquets, and marketing them in the Twin Cities of Minnesota. Other members were raising produce that they sold through farmers' markets or community-supported agriculture arrangements. Although members of the Women's Network did not share a common interest in a particular farming technique, they did share a common identity as women farmers.

Research Perspective

Consistent with an interpretive approach to social science that emphasizes human, lived experience, my research perspective recognizes that people think, interact, and develop lines of action based on their interpretation of situations. As Robert Prus (1992:59) has pointed out, such an interpretive approach requires a method "that is sensitive to the (multi)perspectival, reflective, interactive, and processual aspects of human group life." Accordingly, I chose to use field research methods that are particularly well suited to the task of studying dynamic, small-group situations such as these farmer networks (Frey 1994). The anthropologist Clifford Geertz (1973) described how field research allows for "thick descriptions" of complex situations where interrelated phenomena can be studied simultaneously and as a whole. In the discussion that follows, I first describe the primary research techniques, participant observation, and in-depth interviews used to gather information in the field. Then I discuss my approach to data analysis and interpretation and finally my role as researcher in the networks.

Site Selection

Farmer networks were chosen for study because they constituted concrete settings in which I could explore the role of local production and exchange of knowledge as a social activity occurring at the local level of the sustainable agriculture movement. From among the many sustainable farming networks in Wisconsin, I selected the Ocooch Grazers Network and the Women's Network for the focus of this study for the following reasons.

First, it seemed important to examine a grazing network because the fast-rising popularity of networks focusing on this particular technique made it perhaps the most prevalent type of sustainable agriculture network at the time of this study. Because grazers emphasize the locality of knowledge production and because they must develop site-specific knowledge when they apply the general principles of rotational grazing to their particular farms, a grazing network seemed an ideal place to try to confirm and extend theoretical treatments that have emphasized the role of physical place in the generation of local knowledge for sustainable agriculture. I chose Ocooch from among the many grazing networks primarily for such practical reasons as that one of the network coordinators was interested in facilitating the research and the relative proximity of the group to Madison, making travel easier and less costly.

Second, the sex-segregated character of the Women's Network highlighted the gender dimension in sustainable agriculture and immediately suggested to me the possibility that gender-related concerns might affect the generation and exchange of knowledge. Thus, the Women's Network seemed an ideal place to try to confirm and extend theoretical treatments that have emphasized the role of social location in the generation of local knowledge for sustainable agriculture. Previous research suggests that farm women and men do not necessarily experience farm life or its demands in the same way (Barlett 1993; Whatmore 1988). If that is true for agriculture generally, the Women's Network seemed to offer an opportunity to explore how gender-related experiences might shape the production and exchange of knowledge in the sustainable agriculture movement.

Data Gathering: Participant Observation

Participant observation is a field research method in which the researcher observes and participates in the social group being studied while the action is happening (Singleton, Straits, and Straits 1993). I conducted nearly four hundred hours of such fieldwork, which consisted of an exploratory phase from late fall 1992 until January 1994 and a more intensive and focused phase throughout 1994 until the spring of 1995. I attended thirty-two events sponsored by nine different farmer networks in Wisconsin (including the two focus networks).

In May 1993, I began attending events sponsored by the Ocooch Network during the group's first season of pasture walks, and I attended these sessions until December 1994. This lengthy exposure offered opportunities to observe the pasture walks during two seasons. I went to four pasture walks in 1993 and seven in 1994 from April until October (I attended all but one of the walks held in 1994). In addition, I went to both the 1993 and 1994 winter meetings held during December at a community hall. Similarly, I attended a winter meeting that the Ocooch Network cosponsored with two other nearby grazing networks during January 1994.

Independent of events sponsored by the Ocooch Network, I visited three different farms belonging to five members of the network, and I visited one of these farms twice. These visits – which ranged from several hours to an overnight stay – provided a chance to talk at length with these farmers and to participate in some of the daily tasks of dairy grazing (e.g., moving portable fences to give cattle fresh pasture). For a person who had little knowledge of dairying beforehand, the experiences of carrying a milk bucket to the refrigerated bulk tank in the milk house and of watching farmers go through the rhythmic pace of an evening milking were truly educational. These farmers taught me much of the language of dairying that I needed to know as the project progressed. Moreover, these visits provided an opportunity to get to know some of the farmers in ways that were not possible in the larger gatherings at network events.

My participant observation in the Women's Network began in March 1993 and continued through March 1995. I attended six conferences, four of which were one-day events and two of which involved two days and an overnight stay. In addition, I attended a small meeting held at one woman's farm late in 1993. The purpose of that meeting was to generate specific objectives and outline a grant proposal that they submitted to obtain funding from the Sustainable Agriculture Program. The network also held a few work parties, one of which I was able to attend, and I went to two field days which the Women's Network cosponsored with the Sustainable Farming Network.

As in the Ocooch Grazers Network, I visited four of the farms in the Women's Network independently of network events. I spent a weekend during lambing season to learn and help out on Diane Kaufmann's farm early in my research in May 1993. I also stayed there on one occasion when they were shorthanded and needed help with milking the ewes and other chores and on several other occasions when her family graciously put me up for the night. Two other farmers – Faye Jones and Ann Hansen – were kind enough to host me for the night as well. In addition, Jean Schanen twice gave me a tour of her market gardening operation. All of these visits offered the opportunity to develop rapport with key network members.

To complement participant observation in the two focus networks, my fieldwork included a variety of other events related to sustainable agriculture in Wisconsin. I went to thirteen meetings or conferences where activists and farmers – some of whom were involved in networks – made presentations or participated in workshops. Perhaps the most important among these were the annual Wisconsin Grazing Conference and the Upper Midwest Organic Farming Conference in 1993, 1994, and 1995. I also attended two events – one a field day at a biodynamic dairy farm and the other an Urban-Rural Day – sponsored by the Michael Fields Agricultural Institute, a research and educational organization promoting biodynamic and organic farming systems in the state. In addition, I observed two events in eastern Minnesota sponsored by the Land Stewardship Project, a grassroots group active in sustainable agriculture and other rural issues there. Other important sources of data included newspaper clippings about the networks and their members, personal communications such as letters from network members, documents disseminated at events, newsletters and other mailings to network members, and grant proposals.

In a participant observation study, systematic inquiry and analysis depend on recording complete, accurate, and detailed field notes (Bogdan and Taylor 1975). Field notes are accounts of people, places, activities, and interactions that the notetaker participates in and observes. Typically, I wrote field notes after leaving the field; however, at some events I took cryptic notes in front of people – such as at a meeting where others might also be writing things down – if I felt that could be done discreetly so people would not feel uneasy about my presence. Occasionally, the Women's Network asked me to take notes for the group, thereby enabling me to record actions in more detail. I tried to type my field notes within twenty-four hours of leaving the field; such immediacy is "one of the sacred obligations of field work" (Lareau 1989:206).

Data Gathering: In-Depth Interviews

After two years of participant observation, I conducted twenty-two in-depth interviews during the period from February through early April 1995. The general purpose of the interviews was to discuss with participants certain broad themes that had consistently emerged in my initial analysis of the field notes. My intention was to clarify the similarities and differences between my own and network farmers' interpretations of the issues at hand. Accordingly, unlike a survey interview, which uses standardized questions, I constructed a more general and flexible interview guide that included open-ended questions covering a range of topics (see appendix B for the Ocooch Network and appendix C for the Women's Network).

Having a guide helped to ensure that the same topics would be discussed with each interview participant in both networks. At the same time, however, I tried to retain flexibility and to prevent my own research hypotheses from being the only concern during the interview. As a result, interview questions were not asked in the exact same way or in the same order each time; instead, I tried to fit my questions into the flow of the conversation. In this way, I strived to follow the advice of other researchers who have suggested that it is important to give people a chance to talk about what is on their minds and to relay what is of concern to them (Anderson and Jack 1991; Briggs 1986). When necessary, I tried to probe interview participants so as to clarify, rather than to assume, the meanings of the terms they used.

After I had constructed the guide, two factors influenced the process used to select which network members I would ask to participate in an interview with me. First, I wanted to include the most active network participants because they seemed more likely than those who were less active to be able to comment on the major themes identified in my own observations. A second concern in selecting interview participants was to include a range on those farm and farmer characteristics that seemed to vary considerably within each of the groups, based on my own interpretations and observations of differences. This process – sometimes referred to as "theoretical sampling" (Strauss and Corbin 1990:176–93) or "purposive sampling" (Patton 1990: 169) – enabled me to obtain a diverse and theoretically rich sample.

In the case of the Ocooch Network I identified a core membership that included twenty farmers (fifteen men and five women). From that list, I tried to select potential interview participants who varied on the following characteristics: age, gender, whether or not the farm was certified organic, farm scale, and degree of interest in sustainable agriculture. Applying these criteria, I selected four female and six male farmers whose ages ranged from about thirty to sixty years and one male government agent who had worked with the network and attended the events frequently. Of the farmers in this group, there were three husband and wife teams, but each individual was interviewed alone so as to minimize the possibility that they might influence or silence each other's responses. These couples' interviews took place on the same day, but I was usually able to have privacy with each individual.

In the case of the Women's Network, I identified sixteen core members who had been relatively active. The list was narrowed by looking for a range of characteristics consisting of age, farming experience, farm enterprise and practices, socioeconomic class, degree of involvement in the farming operation, farm scale, and degree of interest in sustainable agriculture. In reviewing the list of potential interviewees, I was able to include women whose ages ranged from the early thirties to early eighties. A few women had lived and worked on farms all their lives, but many others had urban backgrounds that included little or no farming experience.

All of the twenty-two people, eleven in each network, whom I asked to participate in an interview agreed to do so. Each interview lasted anywhere from one and a half to two and a half hours, and all but one of the interviews were conducted at the participant's farm. Each participant signed a consent form that described the purposes of the study and assured the anonymity of the respondent. After reviewing a draft of this study, those individuals interviewed were given the option of whether to be identified in this book by their real name or by a pseudonym.

All of the participants agreed to have the interviews tape-recorded. The tapes were later transcribed in full. Quotations are an important part of depicting the experience and perspectives of actors in the networks. In presenting quotations I have tried to use verbatim language. In instances when the original quotation was unclear, however, I altered the grammar to follow more standard English. Similarly, I sometimes found it necessary to delete words or phrases that were used repeatedly (e.g., "you know") or that interfered with the flow of the quotation as written, rather than spoken, text. As is customary, I have indicated deletions with ellipses.

When selecting the participants for interviews, I held out the possibility that I might conduct additional interviews if that seemed necessary. By the time I had completed all of the interviews, however, I felt that I had reached the level of what Anselm Strauss and Juliet Corbin (1990) call "theoretical saturation," that is, the same major themes kept recurring.

Data Analysis

Participant observation studies based on detailed observations of social activity typically seek to identify, describe, and analyze the concepts and categories that participants use to interpret the world. Such analysis then attempts to link those observations to existing theory about larger social processes, in this case the role of knowledge in social movements. Achieving these goals requires continual analysis of field notes and interview transcripts in an ongoing process of moving back and forth between data and theory.

Analysis in this study involved several different stages. My field notes generally included the first stage of analytic comments, which involved formulating ideas (or hunches) about what I thought was happening in the two networks. I then tried to assess the validity of these interpretations in subsequent fieldwork and, in this way, continually refined this level of analysis over the course of the project.

Just before developing the interview guide, I embarked on a second and more systematic stage of analysis. This consisted of a process similar to what Strauss and Corbin (1990) call open coding. I organized all of the data for each network chronologically and then coded their contents. The coding process involved applying conceptual labels (e.g., "knowledge exchange about pasture ecology," "network as a source of social support," "environmental benefits of alternative practices") to discrete happenings, events, and verbal statements. I used both concepts that participants used to describe their activities and thoughts and concepts that have their origins in the academic literature on local knowledge and on sustainable agriculture. After generating and coding each instance of these concepts in the data, I grouped the concepts together to form larger categories. For example, the concept of "knowledge exchange about pasture ecology" fell under the larger category of "knowledge exchange about farming techniques." Likewise, "network as social support" was part of the larger category "meaning of the network to members." And the concept of "environmental benefits of alternative practices" fell under "ideological exchanges about sustainable agriculture." I recorded each instance of a concept onto coding sheets for each larger category, noting the page number(s) where it could be found and the data source it came from (e.g., field notes, newspaper clippings, newsletter). Throughout the open-coding process, I continually looked for comparisons and contrasts between groups and individuals as an aid toward developing questions and themes of interest that were unanticipated in my earlier theoretical analysis (Strauss and Corbin 1990).

The analysis of participant observation data was used in generating the set of topics and questions included in the in-depth interviews. I tried to expand on the themes that emerged consistently in the analysis and to fill in any identifiable information gaps. To analyze the interviews, I used the same categories and concepts in a coding process similar to that described above.

My Role in the Field

Throughout this project, I constantly tried to define my role as researcher. On many occasions, conducting research was pure joy. I remember sitting around a kitchen table drinking wine and laughing with several members of the Women's Network. And I recall warm afternoons walking across lush green pastures and talking with grazers as Jersey cows chomped away nearby. But field research means much more than these enjoyable experiences. I found it to be an intellectual and emotional challenge as I continually reflected on the research process and as my perspectives changed over the course of the project.

After about a year of attending network events and making contacts in the field, I began to recognize that an ethnographic approach – which by definition takes people's "everyday" interpretations and actions seriously – might contribute something to the broader movement. It seemed that there was much to learn about the way networks operate and that their potential is rarely considered by some in the movement who focus perhaps too exclusively on policy initiatives as a means of social change. Accordingly, I began to see my role as one that would help to articulate the potentials and limitations of the creation and exchange of knowledge that I observed in the networks.

As I focused my thinking on the project, I asked each network to allow me to address the group. I asked the members gathered at those meetings if I could be a participant observer in the group, and I tried to explain what that entailed. I also made a few statements about the purpose of the study. Both groups agreed albeit with minimal discussion about the idea. In the Women's Network, there were a few jokes about how I might "participate" by mucking out the barn, but overall they were very positive and supportive. One woman felt I had contributed a lot by participating in the meetings, even if the research itself didn't end up "benefiting" them directly in the long run. Another woman thought it helped "legitimate" what they were doing. At the Ocooch Network – which was generally not prone to discuss organizational matters very much – people listened to my brief description of the project, but afterward there was no discussion or questions. It is hard to know how to interpret such silence, but I came to believe that it was not necessarily an indication of indifference to the research. The two network coordinators were supportive throughout the project, and they kept me informed on the group's activities. I never heard of any negative reactions to my presence; rather, people were always friendly and seemed willing to converse with me. In addition, over the course of the project, many members of the grazers' network asked me about the research I was doing, and they often claimed to be impressed by my consistent attendance at the network events.

Indeed, as the network members had noticed, ethnography implies that the researcher is engaged for an extended period of time in the field site. Yet how much and in what ways a researcher participates in the situation under study varies across a continuum (Reinharz 1984; Singleton, Straits, and Straits 1993). At one end of the continuum, participation is emphasized and observational activities are either somewhat concealed or completely covert. At the other end, the observer minimizes interaction with members of the field site and considers observational activities overriding.

In general, I would characterize my role in the field as closer to the active participant end of the continuum, although I was never covert about my purposes. Like others who have done participant observation, I found ways to participate in the network during the course of the project, such as by passing along information that I knew members would be interested in or by running an errand in Madison. And, unlike the researcher who emphasizes observation over participation, I developed relationships with people that I met, offered my opinion occasionally, and tried to minimize the amount of notetaking that I did in the field. In certain ways, it was easier for me to be a slightly more active participant in the Women's Network than it was in the Ocooch Grazers. Perhaps this was because – as a woman – I was somewhat more comfortable in a space dominated by women than I was in the grazers' network, which tended to be male-dominated. Similarly, I had more opportunities to spend informal time getting to know several farmers in the Women's Network. In contrast, the more frequent – but relatively short – Ocooch pasture walks were often over just before milking time, and people did not usually linger too long afterward for informal conversation.

Although I was engaged in the activities of the networks over a considerable period of time, I was also acutely aware that I was an outsider and somewhat uneasy about my role as researcher. For example, early on a farmer commented on liking the premise of my research but also expressed feeling somewhat like a "bug under a microscope." Although that person and I have since discussed my research and our mutual uneasiness, the comment haunted me through much of the project. I thought much about my own ethical considerations in not wanting to exploit those whom I had come to know in these settings. Accordingly, I sympathized with Annette Lareau (1989:207), who wrote of her own experiences: "There was a lurking anxiety about the field work: Was it going right? What was I doing? How did people feel about me? Was I stepping on people's toes? What should I do next? – and this anxiety was tiring." Overall, in my view, such questions and struggles constitute a healthy, reflective process that can sensitize the researcher to the advantages and disadvantages that are an integral part of field research.

Of course, the researcher's role does not end when one leaves the field. Rather, the final role a participant observer plays is in writing about the social groups studied. Scholars from a variety of disciplinary perspectives are engaged in critical conversations about the ways in which researchers construct the narratives we call social science (e.g., Borland 1991; Fine and Vanderslice 1992; Haraway 1988; Lather 1991). It is now well recognized that writing about what is learned in the field is itself an interpretive act. Clifford Geertz (1973:9) was one of the first to recognize that these narratives are "really our own constructions of other people's constructions of what they and their compatriots are up to." Similarly, I have identified what I consider to be the prominent themes that emerged in my observations of the networks and other social activity in Wisconsin's sustainable agriculture movement. And I recognize that what follows is my own interpretation and not a straightforward reading of reality. Nonetheless, I hope that those involved in the sustainable farming networks will find much in my interpretations and analysis that reflects and informs their own knowledge and experience.

During the course of the last ten years, sustainable farmer networks in Wisconsin have developed considerably. In general, networks have moved away from the funding and support provided by the Sustainable Agriculture Program as a result of diminishing funds and changing priorities in that government program. Although this institutional mechanism was a key element in establishing the farmer network system, other factors apparently contribute to the continuing formation and maintenance of networks. This study was designed to examine the principal activities and functions of the farmer networks, with a particular focus on the role of knowledge exchange. In the next chapter, I describe and analyze the Ocooch Grazers Network before turning in chapter 5 to the Women's Network.

FOUR

Dairy Heretics and the Exchange
of Local Knowledge

In the late 1980s and early 1990s some dairy farmers in Wisconsin began to experiment with milk production based on intensive rotational grazing as a technological alternative to year-round feeding of stored forage to livestock kept in confinement. As these new "grass farmers" abandoned many tenets of the conventional system and thereby committed what the Wisconsin journalist Joel McNair (1992a) called "dairy heresy," they turned to one another for much of the knowledge they needed and simultaneously asserted the validity and utility of farmer-generated knowledge. They did not necessarily do so, however, because they believed that institutionalized agricultural science *could not* help them (if the relevant research were done) but simply because it *had not* helped them. The public and private agricultural research system did not have much useful information to offer Wisconsin dairy farmers exploring rotational grazing. An agronomist at the University of Wisconsin stated the problem succinctly: "I get asked more questions that I can't answer than I can answer about rotational grazing. . . . There's been no research done."

Significantly, both public and private researchers have begun to respond to advocates of sustainable agriculture and to the expanding interest in intensive rotational grazing among Wisconsin dairy farmers (Gale-Sinex 1994; Rust et al. 1995). Although the new research initiatives are promising, most grazers have recognized that, as one grass farmer put it, "it takes a long time to turn the supertanker around" (quoted in McNair 1992c:14). Instead of turning to agricultural research institutions, the first rotational grazers sought information and support mainly from sustainable farming networks that many of these same farmers had been involved in developing. And new networks were eventually formed such as the Ocooch Grazers Network. As the following description and analysis illustrates, members of the Ocooch Grazers Network pursued a distinctive approach to the generation and exchange of knowledge that was only tangentially related to the changes occurring in research institutions.

This chapter begins with an organizational profile detailing how this network's structure constitutes a "social movement community" (Buechler 1990; Stoecker 1995). The second section focuses on my finding that, although grass farmers constructively learned the technique's basic principles from a variety of sources, for network members the practice of rotational grazing was profoundly contingent on the adaptation of those principles to place and the personal development of local knowledge. In the third section, contrary to interpretations that emphasize the deeply personal, tacit character of local knowledge in agriculture, the analysis I put forth suggests that local knowledge was being meaningfully shared or socialized among network members. Indeed, the primary function of the network was the exchange of local, technical knowledge, as members pursued a collective effort to answer questions about how to graze.

The fourth section of this chapter explores the ways in which the exchange of practical knowledge about *how to graze* was complemented by the creation and exchange of values, ideas, and beliefs about *why to graze*. While many of the topics of knowledge exchange within the Ocooch Grazers Network were practically and materially important to the grazers as producers, network members simultaneously challenged certain prevailing assumptions underlying conventional agriculture. As in all processes of knowledge creation, these interpretive frameworks influenced the more technical questions that were asked and the answers that were developed. Finally, in the last section I analyze how the network was not only an important source of information and ideas but also an important source of support for the farmers involved. Indeed, network members began to find themselves disconnected from their neighbors who continued to farm conventionally. Although there were significant differences among the members, particularly in the extent to which they identified with the broader sustainable agriculture movement, their participation in the network provided social support for their unconventional practices, and that support constituted another important function of this social movement community.

Organizational Profile

Social movements give rise to a variety of organizational forms that may be thought of as spanning a continuum from tightly to loosely structured (McAdam, McCarthy, and Zald 1996). At one end lies formal, often bureaucratic, organizations that are organized solely to advance a movement's goals. In the United States, such organizations typically establish by-laws and constitutions and hire professional staff. In contrast, farmer-to-farmer networks, such as the Ocooch Grazers Network, fall closer to the other end of the continuum. Unlike political parties, unions, or professionalized organizations, these loose associations tend to focus on local and specific issues rather than explicitly trying to advance the goals of the wider movement. With relatively simple arrangements and informal patterns that characterize how the group functions, such associations are typified by a fluctuating membership base, a minimum of hierarchy, and a loose federation with other groups.

To better understand these informal groups within social movements, Steven Buechler (1990:42) introduced the concept of "social movement community" to designate "informal networks of politicized individuals with fluid boundaries, flexible leadership structures, and malleable divisions of labor." Buechler (1990:43) proposed limiting the term to only those "communities that identify their goals with the preferences of a social movement and attempt to implement those goals." This general concept of a social movement community is useful for describing the organizational structure of the farmer networks. Like the geographically localized neighborhood movements studied by Randy Stoecker (1995), however, we shall see that a farmer network community includes people who participate for reasons that are not explicitly political and that it can contain both activists and nonactivists who are networked to each other. For this reason, Stoecker (1995:112) suggested broadening the definition of movement community to "intersecting social networks in which a collective of movement members are embedded" and in which people who do not explicitly identify with the movement may be involved. Since February 1993, when Mike Cannell and Jim Brown organized the Ocooch Grazers Network's first meeting, the group has maintained a very simple structure that resembles that of a social movement community.

Membership

Membership in the network during the period I studied was open and fluid. Technically, all those who asked to have their names on the network's mailing list were considered to be members regardless of occupation or geographic location. The original organizers intended the network to be for dairy farmers in Vernon and Richland Counties, an area of the state that appears to represent the highest levels of adoption of rotational grazing among dairy farmers (Jackson-Smith et al. 1996). But only 55 of the 110 individuals and families on the 1994 mailing list (50 percent) were from those two counties. Another 34 lived nearby in five bordering counties (Crawford, Grant,

Juneau, Monroe, and Sauk), and the remaining 21 members were scattered around the state. The vast majority of those on the mailing list were farmers, but at least 11 percent of those listed were not engaged in farming and were connected only indirectly to the practice of rotational grazing, including a local banker, an agricultural economics professor at a local technical college, three agricultural extension agents, a district conservationist with the Natural Resources Conservation Service (NRCS), a staff person with the Wisconsin Rural Development Center, a local veterinarian, a university researcher, an agricultural journalist, a representative of a cattle breeding company, and a production consultant for a major dairy processor.

Maintaining one's membership in the group was simple. Each year the network coordinators sent out a short form asking members if they wanted to remain on the mailing list. No membership fee was involved, and the only monetary cost required for maintaining membership was the postage to return the form to Mike Cannell, who kept the mailing list on his personal computer. There was always some turnover of the membership, according to Mike: "We have probably 50 percent of the people who phase out every year. You know, it gets up to 150 [members] by fall, and then it drops to 70 or 75. And then by the next fall, it's back up to 100; so there's a lot of turnover."

As with most groups, the level of participation in the network varied among the members. Some people's involvement was limited to contact through the periodic mailings. Other members attended events only two or three times a season. In addition, nonmembers came to some network events, all of which were open to the public and were often listed in a calendar published in a statewide weekly agricultural newspaper. As a result, participation at one event might draw people who came only once, such as the time when six people came some 120 miles across the state to attend one pasture walk because no network existed yet in their area.

The focus of this study was on the active membership; therefore, I made no systematic attempt to find out the reasons why some members came only occasionally or not at all. Mike offered an explanation:

Somebody may go to a pasture walk because somebody said they should, and they might never come to another pasture walk. They might say, "Well this is a stupid idea." And when you go look at some pastures, you should think it's a stupid idea because everybody is in this transition stage where it's not picture perfect. . . . I've been at it eight years, and you still go out in my pastures, and they are not mature pastures. . . . When people come to experience what you're doing, there's a certain level of expectation. And if you don't meet it, they are liable to say, "This guy's crazy" and just go home and get on the green feed chopper and go out and do whatever they were doing before.

This explanation suggests that one reason some existing or potential members may have been dissuaded from participating is that the successful transition to this unconventional technique was a slow process, as was the development of a sufficient level of knowledge to understand such a transition.

In contrast to those who came and went, for some twenty farmers network membership meant much more than receiving mailings or attending an occasional event. This group of fifteen men and five women constituted a core that regularly attended pasture walks or other network-sponsored events during the two years of this study. Typically, these core members were the ones who hosted a pasture walk at their farms. They encountered some indirect costs of membership that were not required of people who attended less often, such as transportation to events and, for the farmers who hosted a pasture walk, provision of snacks for the group. Still, these core members gleaned some significant benefits from their participation which offset such costs. It is the sustained interaction among these core members that is of interest here.

Leadership

The network's informal structure included the minimum level of hierarchy necessary for the group to function reliably and consistently. At the network's inception, Mike and Jim expressed their willingness to serve as unpaid coordinators. Although the task of network coordination was the only explicit leadership position in the group, there was little clarity about what this duty entailed. Reflecting the network's casual approach, decisions about the coordinators' roles were made largely by default. In other words, most decisions were left up to the coordinators themselves, and the active membership never challenged them, perhaps connoting implicit agreement with the direction the group was being taken.

The coordinators assumed responsibility for specific tasks and divided the work between themselves. In general, Mike was in charge of scheduling network events, making sure participants were informed about the time and location of events, and acting as group facilitator on occasion. Jim focused primarily on passing along any information he deemed to be of potential interest (other than scheduling) to the membership. They each contributed to preparation of pasture walk reminders and mailings sent out monthly during the growing season and when necessary during the winter months. Mike communicated specific logistical information and often added his own thoughts on grazing-related topics. Jim's messages in the pasture walk reminders usually took the form of a letter summarizing ideas learned at the previous pasture walk and conveying new practical ideas he had come across or tried out himself.

The coordinators encouraged input from members. There were two primary mechanisms for members to give feedback about how the group was operating. Typically, the form mailed out annually on which one could request to remain on the mailing list included several questions designed, as Mike wrote in an introduction, "to give everyone a chance to comment and make suggestions as to who we are as a group and how you feel the group is doing." Similarly, at the two annual winter meetings I attended, Mike asked if the members present wanted any changes in how the group was operating. A few people suggested ideas for topics they would like to see the group discuss (e.g., pasture plants that might be poisonous to cattle) or expressed an interest in having the pasture walks held on a different day of the week. But in general, the members engaged in little discussion about the mission of the network and whether it was adequately meeting their needs. While silence on this topic did not necessarily imply complete satisfaction with the group and its leadership, the members I interviewed were quite content.

Because the coordinators were not elected, it was unclear how long they would continue to serve. It was mainly up to the coordinators themselves to decide; the members never raised it as an issue. In part, the coordinators continued to assume their responsibilities because they enjoyed those responsibilities and were willing to do the work. But they did express concern that the group not become associated solely with them just because they were the ones who started it. Mike, in particular, hoped the network would be concerned about more than rotational grazing per se so that it would help "to create a base of rural leadership." As he explained:

I don't want to become a dictator in the grazing network. I don't want to become a guru. I don't want to become this father figurehead, this guy that's been running this grazing network for the last fifteen years. . . . I want the grazing network to be strong enough to stand on its own, and I want people to cycle in and be active in running it, and then cycle back out. I'm in favor of term limits when it comes to the grazing network. I want to share the responsibility, as well as the authority and the satisfaction. And it's satisfying to me to see what it does. I thoroughly enjoy using the computer, so I enjoy very much creating these little pasture walk reminders. I enjoy doing it, which is one reason why I haven't looked very hard for somebody to take my place.

But I still feel that if the network is going to be strong, there has to be an equity of involvement. It can't be this two or three people up here and then everybody else dragging along, not knowing what's going on. So we're working on that. That's obviously not related to the fact that it's a grazing network. It's related to the fact that it's an organization, regardless of what the organization is organized for.

Each year during the winter meeting, Mike mentioned to the group that he would like to see others take over some of the coordinating tasks and that he would not be willing to do it "forever." After two years, he began asking other members whom he thought were "administratively capable" if they would be willing to take over for him. But he never pushed the issue until he really wanted to give up the responsibility, and Sally and Tom McMahon agreed to take the position in 1996 (after I left the field). Jim continued in his role after Mike stepped down. Despite Mike's good intentions of developing what he called a "democratic" organization, the leadership recruitment and development processes were never formalized in ways that led to an equal amount of leadership input from all members.

Activities

The Ocooch Grazers Network used two activities, pasture walks and winter meetings, to facilitate farmers' exchange of knowledge about rotational grazing. I will sketch these events only briefly here, but more specifics will be provided in subsequent sections where knowledge exchange is described in detail.

Pasture walks were scheduled to take place on a different member's farm once a month from April until November. At these very informal affairs, the coordinators did not communicate any agenda to be followed closely, but a pattern did develop. Participants typically arrived at the host farm at one o'clock in the afternoon. While waiting for the pasture walk to begin, they would chat in small groups, and the young children accompanying their parents would go off and play together. Farmers and other participants dressed casually in clothes appropriate to walking through pastures, manure piles, and often muddy lanes.

The event would begin with Mike or Jim getting people's attention, and participants gathered together in something of a circle. Mike then asked people to introduce themselves to the group. Participants stated their names, where they lived, the number and breed of cows that they milked, and perhaps how many seasons they had been practicing intensive rotational grazing. Almost like a roll call, these introductions were short, and participants often seemed unwilling to say much about themselves, perhaps out of shyness or simple familiarity with the others present. The women at pasture walks were particularly brief in describing themselves, sometimes saying, "I'm with him," after her partner described their operation. Or occasionally a woman was skipped over. Whenever this happened, Mike would stop the flow of introductions, ask the woman to introduce herself, and remark that it was important to include everyone.

After the introductions, the coordinators made announcements and on occasion asked the group to make a decision about such things as access to the mailing list. Jim, who had a penchant for inventing gadgets, would often take a moment to do small demonstrations of his innovations or share tips. Next, the host farmer(s) would talk briefly about the farm, its history, and their views on rotational grazing. Typically this presentation was done by the man in the family, although on three occasions I observed a husband and wife participating equally in describing their farming operation and answering questions. (Only once during the two years of this study did a woman make the presentation and lead the walk entirely, apparently because her husband was not involved in their farm, but I was unable to attend that pasture walk.)

Next, the group would head out into the pastures, where the outdoor setting influenced the dynamics. Without walls or tables to limit group movement, people walked about freely and with whomever they chose. Some remarked that they liked to move around between the various conversations to see what different people were discussing. Rarely did one person hold the attention of the group as it moved, but occasionally people gathered in a spot, and specific issues were raised for discussion. After walking through the pastures, and occasionally through the barn, the group would return to the central location where it first gathered. People often stayed for a while and enjoyed informal discussions and snacks, which invariably included milk from the host farm. Slowly, the crowd would dissipate around three o'clock, leaving enough time for these dairy farmers to go home for the evening milking.

Although the pasture walks were the cornerstone of the network's activity, winter meetings were held in November or December to conclude the grazing season and plan for the next one. Compared with the pasture walks, these meetings at a community hall involved somewhat more formal presentations focusing on particular topics of interest to the members. For instance, after sharing a potluck meal, the 1993 winter meeting featured five farmers, some of whom were members of Ocooch and others who were from neighboring counties. In a series of hour-long sessions, these farmers, all of whom had been practicing grazing for some time, spoke to the group about topics that illustrated what Mike Cannell called "the progression of thought" that grazers went through as they tried to maximize the potential for grass-based dairying in their operations. For the 1994 winter meeting, the coordinators invited a plant breeder from the University of Wisconsin to answer questions from the members on various pasture plant species. The winter meetings also provided an opportunity to conduct network business. The coordinators set the schedule for the next season by asking members – especially those who had not already done so – to volunteer to host pasture walks the following year. For months that were not filled, the coordinators later recruited other members to host walks and complete the schedule.

In addition to their own winter meeting, Ocooch also cosponsored a Tri-County Rendezvous with two other networks from neighboring counties. Held in January and billed as "a pasture walk in the high school," these events featured several speakers who made formal presentations to a fairly large audience of about 150 people.

Linkages with University and Government Agencies

From Mike's previous experiences working with other rural and farm advocacy organizations, he knew that running an organization requires money. The network's expenses were principally the cost of mailings to the membership. So at the group's inception, when a local University of Wisconsin extension (UWEX) agent and a local district conservationist with the U.S. Natural Resources Conservation Service offered to help with the network, Mike accepted. These two local offices alternated responsibility for mailing out the pasture walk reminders and other announcements to the network members, and they added a brief cover letter to each one. In establishing this relationship with these agencies, however, Mike was clear that he wanted the network to be "independent of typical university control," and he set some boundaries:

I said to myself, "If I can get them to bear the financial burden, I will bear the administrative burden." I said to them, "The mailing list is going to be at my place and my computer, and that's the only mailing list." . . . I send them the labels, and they put the label on the envelope. . . . So we try to keep thorough control over it, not because we're control addicts but because we don't want it to be diluted with other stuff that extension might want to . . . distribute. We wanted it to be more pure than that. We knew . . . that a lot of people who would consider going to grazing had a certain animosity toward the university anyway. I mean there's a certain philosophical mind-set here. If I'm a conventional, concentrated dairy farmer with all this equipment, and I'm tied in close with the university extension person and these researchers, I'm not apt to think about going grazing, see. So if you have a farmer that says, "By golly I think I'll go grazing," that farmer probably already is willing to push the university aside and decide to do this. So I knew that in order to maintain credibility among that type of person, I had to keep the university at bay. . . . I could use them, but I had to keep them under control, so to speak.

In other words, instead of relying on existing agricultural institutions for information, Mike and Jim relied on them in another way.

Both the UWEX agent and the NRCS district conservationist who maintained ongoing relationships with the group seemed willing to help the network achieve its goals. Of the two, Jim Radke, the NRCS district conservationist, participated in the network events more regularly. He wanted to see intensive rotational grazing more widely adopted in his county because he recognized its potential benefits for protecting water and for soil conservation. But he said that he did not assume much leadership in the network for the following reasons:

I wanted to try to get other people to know about this system, and yet I don't have time to spend on intensive rotational grazing. And I really liked the networks because people didn't call me. They called the networks. And if someone calls, all I have to do is say, "Here, call one of these people from this network." And so I saw an opportunity to help them because I know how these groups are, they don't have any money to begin with. And I told them that, if they wanted, I could help them with the mailing. All that I had to do was put a cover letter on any correspondence they had about their meetings. Give me a mailing list, and I would send out the mailing list for them. . . .

My goal was to try to get more farmers looking at the system and to help them get organized. But I've never really even helped the Ocooch Grazers with anything except send out those mailings intentionally. . . . We don't have the time. We're getting cut back. We're going to lose a person in April, right at the heart of our spring layout system, and the last thing in the world I can afford to do is sit down and talk to three people about intensive rotational grazing. . . . So it's kind of funny. I want to see the system work, and yet I know I don't have time to invest into it because the amount of work that I have to do has been pretty well dictated by Congress, and they do not include this kind of work right now. . . . I don't have a role in the network, and I don't want a role in the network. I want to see the networks become as independent and self-functioning as possible. I guess, if I have a role, it's to accomplish that with spending as little time as possible.

Clearly, while the grazers pursued their personal and collective goals, they did not act in social isolation.

Adapting Principles to Place

Several analysts and advocates have stressed that experientially based, local knowledge can be a valuable resource in the creation of sustainable agricultural systems. Local knowledge is the practical skill that develops with mindful attention to the unique, yet shifting, social and physical features of a locality and that is fundamentally tied to direct, personal experience of a particular place or activity. One of the purposes of this study was to explore the role of local knowledge in a particular embodiment of the wider sustainable agriculture movement. In other words, what is of interest here is production of local knowledge with respect to the particular activity of grass-based dairying in a specific physical setting.

Members of the Ocooch Grazers Network, like other dairy grazers (Hassanein and Kloppenburg 1995), were engaged in a fundamental reversal of many established patterns of technological development in Wisconsin dairying. Conventional dairying focuses on maximizing milk production by feeding stored forage to cows confined primarily to the barn and surrounding area. Typically, on such farms most land is used for producing crops such as corn, alfalfa, and soybeans, and dairy cattle are fed year-round by the farm operator. Of particular benefit to agribusinesses is that the conventional production process requires major capital outlays to cover the costs of the fuel, equipment, silos, and other inputs needed to produce and store feed, to move the feed to the cows, and to remove and redistribute manure (Murphy and Kunkel 1993).

Conversely, rotational grazers manage permanent pastures so that cows harvest forage and spread manure themselves during as much of the year as possible. Relying on pasture to provide the bulk of the cows' rations during the so-called grass season allows dairy grazers to reduce corn, alfalfa, and other field cropping significantly. Some farmers eliminate these crops entirely and purchase winter feed elsewhere. Thus much, if not all, of a grazer's land is taken out of monocultural production and put into perennial polycultures of legumes, grasses, and other pasture plants. In describing the shift from the conventional system of feeding dairy cattle to rotational grazing, one network member said plainly: "For years, I fought and borrowed to bring feed to the cow. Then I finally figured out that I could bring the cow to the grass."

Although this technological reversal sounds simple, questioning conventional wisdom and actually learning how to "bring the cow to the grass" proved to be difficult tasks for members of the Ocooch Grazers Network. The technology associated with rotational grazing (principally high-quality permanent and temporary electric fence systems) is relatively straightforward compared to the increasingly complicated mix of technologies associated with confinement dairying (e.g., Total Mixed Ration machines, bovine growth hormone). What is critical for successful grazing, however, is not so much buying or understanding how to use new tools but rather learning new ways of seeing and thinking. Spending time with grazers and attending Ocooch pasture walks, I frequently heard farmers stress the importance of "experience" and "observation." For example, in response to one farmer's question about how much pastureland dairy cows should be given each day, another network member replied, "You only know by experience." And at pasture walks, Jim Brown often made remarks such as this one: "You have to look ahead and determine when to graze and that requires experience and observation. There aren't any rules or regulations. You gotta look at the pasture. You can't look at a calendar and know what to do. The whole thing is observation."

This emphasis on developing one's skill through experience and observation led many grazers to say that there are no "recipes," "magic formulas," or "pat answers from textbooks" when it comes to grazing. Most members, however, acknowledged during interviews that several key principles undergird the technique. Rotational grazing hinges on the basic idea that pasture plants must have a chance to photosynthesize and replenish energy reserves after each grazing. Grass farmers accomplish this by dividing land into small areas and rotating animals through these paddocks according to the time necessary for proper regrowth and recovery of the pasture plants. Sufficient residue has to be left behind in the pasture after the cows are moved on to the next paddock so that pasture plants can regrow easily. By employing short grazing periods with high stocking density on these paddocks, a grazer tries to ensure that nutritious and palatable forage is available for the animal and that plants are not overgrazed. Thus through proper rationing of the forage and providing enough time for plant recovery, grazers seek to match the nutritional needs of the animals with the growth rate of the pastures.

Managing such a complex system demands skill, flexibility, ingenuity, and profound attention to detail. Members of the Ocooch Grazers stressed that successful grazing requires observing and interpreting the constant flow of signals from a dynamic and variable production system in order to know when a pasture is ready to be grazed. For example, during what grazers refer to as the "spring flush," pastures grow fast. As a result, cattle have to be moved through the rotation quickly to prevent the pastures from exploding with growth and the quality of the forage from deteriorating. Conversely, late-season grazing demands much slower rotations when grass growth has diminished and often requires incorporating more land into the rotation so that the rest periods can be lengthened and sufficient regrowth can occur. In addition, Ocooch grazers spoke of trying to create a "stepped condition" in the paddocks so that each paddock would be at a different stage of growth, and not all paddocks would be ready to be grazed at the same time. Several Ocooch grazers also experimented with extending the grazing season by "stockpiling" forages in the pastures to be grazed in late winter or early the following spring. These and other management judgments were made within variable regimes of weather, soil, pests, diseases, and socioeconomic factors.

Rather than working as hard as they could or relying on technology to overcome problems, these new dairy grazers said they needed to develop a sophisticated "grass eye" so as to match the nutritional needs of their cattle with the growth rate of the pastures. According to network members, one of the first skills a grazer needed was the ability to decide how much forage was available in the pasture for the number of cows being milked at a given time. Joanie Brown described the process through which she learned how much pasture to feed the cows in this way:

You go out there, and if you've got ten inches of beautiful clover and alfalfa and timothy and all this mixed in the pasture, and you give that to them in the morning. You figure out so many steps is what we do [i.e., pace off the size of the paddock]. Anyway, you figure out what you think they're going to eat for that feeding. And then if you go out there the next feeding – we move the fence twice a day – and that's just chewed down to nothing, then you know then you didn't give them enough. So then you give them more the next feeding. Or if you go out there, and they've just trampled it and there's so much left, you know I gave them too much. So I'll have to cut back the next feeding. You learn what to do. And you can go by their manure too. If they're getting far too much protein, their manure is just like water. It just goes right through them, so then you've got to kind of change their diet. But [grazing] is not just going out and putting them into pasture. There's so much involved.

Another important skill that grazers said they developed was the ability to understand how certain pasture species performed in a particular grazing environment and how to manage those species according to their characteristics. Orchard grass was one pasture plant that several members had a difficult time working with, but Mike Cannell described the process by which he learned how to manage it on his farm over the course of several years:

I have five acres of orchard grass that works just fine because I've learned how. The first two years, when I put the cows out there, I couldn't see the cows. The orchard grass was that tall and headed out, and I said, "This isn't working." And then I realized it grows so very fast in the spring. So now when I put my cows out, I put them out there and just beat it to death, and go around the rotation. And then bring them back and beat it to death, and then I'll send my heifers out there and they're going to beat it to death for two or three days. But I make sure I keep that plant down here. I never let it get away from me.

Mike Cannell summed up this process of developing the grass eye: "The grass growth rate demands forethought and that's the part where there's no recipe. That's the part that you just have to gain experience."

Gaining experience, however, can be a painful process, as several members pointed out. For instance, Sally McMahon, who moved to a rural area with no background in farming and eventually began a dairy farm with her husband, Tom, explained that when they began grazing, several of their cattle bloated. Bloat occurs when a foam forms in the rumen as a result of a complex interaction of animal, plant, and microbiological factors; this foam causes pressure and swelling in the rumen, which then presses against the lungs and prevents breathing (Murphy 1991). The animal dies if not treated quickly. Bloat can occur when cattle graze on lush pastures with succulent legumes. According to Sally:

In the early days, we used to have a problem with bloat. Knock on wood, but we don't any more. I think we've learned the ins and outs about how to bloat a cow and how to try not to bloat a cow. But the learning curve in the beginning can be a little expensive, if you bloat too many cows. . . . In the spring and the fall when the grass gets frosted, it has a higher capability of bloating the animal. After milking you send them out to the next paddock, and in the spring and the fall that can be real dangerous if you send them out too hungry or if it's too high in alfalfa. There's a learning curve in the beginning to figure out how to manage pastures. . . . We didn't kill anybody last year to bloat. We have in years past.

Another example of the sometimes difficult process of learning how to graze came from Joanie Brown. She initially introduced the idea of rotational grazing to her husband, Jim, when she found and brought home from a garage sale some old farming magazines that described how the technique was used in the 1950s. Even so, she was very skeptical about going to grazing at first, and when they began to experiment with it on their farm in the late 1980s, trouble hit:

When we first started grazing, we were told don't give them any feed in the barn, they don't need any even to bring them in. . . . We tried that, and boy our cows were in bad shape that first year. I mean really, they were so thin. They didn't milk. I think our herd average was like ten pounds a cow per day. Really, it was bad, and I was so mad. I mean I was totally against grazing. I didn't want nothing to do with it that year. We were just about in divorce court because I said: "We're not going to do this again, never. You can't graze, you can't do it." Anyway, so he studied more into it, and then the next year we started giving them grain just to get them in. You know they each got a scoop of feed, which is like a pound and a half. . . . Well, then they did better, but they still weren't in good production, and breeding was bad. So then I said, "Well, that's it. Now we tried it two years, and the cows are not doing well." So then the next year, we supplemented them with round bales we put out there for them to eat, and we gave them the feed in the barn and figured out what they needed and their minerals, made sure they had it. Now we not only give them what they need in the barn, if the pastures ain't as good as they should be, we'll also give them hay in the barn, even if the pasture is really good, [because of] the protein. They get the poops from it, so then you still have to give them dry hay. So you have to feed them according to what's out there. You can't just turn them out. We learned that the hard way.

After learning the "hard way," Joanie came to appreciate the fact they had switched to grazing, saying, "If you've mastered it, boy, you can't beat it." Thus, though grazers stressed that they learned by doing, it was also clear that experience could be a harsh teacher.

According to network members, the development of their personal grazing skills was an ongoing process. Michael Hall described it: "Once you start doing something you really become knowledgeable, and it's a never-ending process because you never have two years in a row that are the same. . . . Things are always changing, and you can't rely on always doing things the same." As a result, network members reported that they had to make frequent assessments and be able to respond to changing conditions. As Glenn Scoville put it:

You've got to keep the feed uniform, and this is probably the hardest thing to do. . . . Probably every three or four days, I'll take my four-wheeler and go clear around the farm and look at every paddock and kind of make a judgment of how things are going. I think the first year, I didn't do that near enough. This is something that I think a person has got to learn – it doesn't take very long, maybe fifteen or twenty minutes – judge how it's growing and make a judgment then of what you're going to do a few days down the line.

In other words, local knowledge is the product of local observation. Such local observation led network members to work within the natural parameters of their particular landscape. For example, in describing how her family set up their pasturing system, Vicki Braun said: "We had to work around the terrain. So for our steepest hills, we didn't divide them up according to size and the number of cows we had. We just divided them according to where we could get the posts in the ground because of the rocks and because of the hills. So we actually followed natural borders in a sense." While this permanent fencing remained, they continually made additional paddock divisions: "We end up changing them every year, in terms of dividing, subdividing, and stuff because of the growth and the weather, and it really varies a lot." For network member John Hills, being attentive to the landscape and responding to the changes that occur there meant that "you're continually thinking of better ways to do it. . . . It's very much of a thinking process."

This emphasis on grazing as a "thinking process" applied to a specific place contrasts with the characterization of conventional agricultural practices as a set of prescriptions that can be applied anywhere without a detailed understanding of the local agroecosystem (Berry 1984). For this reason, artist and dairy farmer Tom McMahon believed that "grazing isn't a science, it's an art." He explained his perception of the difference:

In a science things happen the same way all the time. I mean A plus B always equals C. In rotational grazing, A plus B might be C, it might be D, it might be E, depending upon a bunch of other factors that you can or cannot control and may or may not even see. So what you have to do to graze is you have to walk out the door every single day and think and look at what's out there and make a reactive judgment. And you're reacting to the situation every day instead of creating a situation based on your preconceived ideas. So you go out every day, and you look at the grass, you look at the cows, you look at the weather, and you make a decision. . . . You don't do it the same every day. You're constantly reacting to the surface or to the environment. And people that really paint, that's how they paint. They don't have a thing in their head that they're putting on the canvas. The canvas is alive, and you're reacting to it all the time. And when you make that first mark, it changes everything. . . . [You have] to understand the variables and know how to react to them.

This continual process of action and reflection meant that local knowledge for rotational grazing had an intuitive feel that developed over time. Mike Cannell summarized the knowledge grazers must acquire:

An intuitive understanding of relationships among multiple variables, this is the intellectual plane that we as grazers must reach; the "spot" of awareness, or the point at which we transcend the reality of the multiple variables and acquire the wisdom which allows us to know the impacts of the interrelationships of those variables; to be able to "see" the weighted average of the forces brought to bear on the grazing event by the various factors at play.

This ability constitutes a *local* knowledge, one that is intimately tied to the performance of a specific activity in a particular place and time.

Thus for network members, if the principles of intensive rotational grazing were relatively simple in concept, they were also deeply complex in application. Even André Voisin, the French farmer and scientist who first elaborated many of these ideas, recognized this complexity. In his book *Grass Productivity*, originally published in 1959 and reprinted in 1988, Voisin presented what he referred to as the "laws" of "rational grazing." Nevertheless, throughout his large volume, he repeated his belief that an irreducible component of grazing was the province of the farmer and not of the scientist. Voisin wrote ([1959] 1988:178): "It is not a case of rigidly obeying figures: one must follow the grass. One has no right to say: so many days after grazing at such and such a time, I will start grazing again. One must look for the plots that are ready for grazing and graze them. Figures are only guides: in the end it is the eye of the grazier that decides. . . . The grass commands; the eye of the grazier follows in its train, ready to receive its orders."

In addition to Voisin's book, there were several other important sources of how-to-graze information used by members of the Ocooch Grazers Network, but those sources too usually led to an emphasis on local experience and knowledge. One such source is New Zealand farmers. Although the U.S. dairy sector pursued the development of confinement feeding, dairy specialists and farmers in some other parts of the world did adopt Voisin's ideas and improved on them over the years. New Zealand is the most notable example of an agricultural system that relies almost exclusively on permanent pastures. New Zealand farmers feed about the same number of cattle as there are dairy cows in the entire United States, seven times more sheep, plus one million each of deer and goats. But they do it on a pasture area the size of Wisconsin and without grain supplements, which are very costly there in comparison to the United States (Murphy 1991:17). The new dairy grazers in the Upper Midwest found New Zealanders' tools (e.g., certain kinds of electric fencing technology) and some of their management techniques to be very useful. Grazers from Wisconsin reported learning from their New Zealand counterparts through a variety of means. Ocooch Network coordinator Jim Brown subscribed to and read a dairy publication from New Zealand. A few network members had even traveled there to visit farms, subsequently speaking publicly and writing about their experiences in popular publications (e.g., Pulvermacher 1993). In addition, New Zealanders have come to the annual Wisconsin Grazing Conference to speak, and those speeches were often reported in the agricultural press.

In learning from their New Zealand counterparts, members of the Ocooch Grazers often stressed that such knowledge cannot be mechanically transferred and applied in Wisconsin. For example, network member George Ball met with a grazing consultant who now works in Wisconsin after spending much time in New Zealand. The consultant showed George some slides that included pictures of cows grazing through the snow there and suggested that grazers in Wisconsin could similarly extend their grazing season. Afterward, during a pasture walk, George related the story of the consultant's pictures when the discussion turned to the idea of reducing winter confinement feeding costs and leaving cows out on pasture into the winter months. Pointing out that "New Zealand snow is not Wisconsin snow," George expressed his frustration at the consultant's unwillingness to "adapt his New Zealand style to Wisconsin," saying emphatically, "We can listen to those guys, but we got to use our heads." Another farmer agreed, adding: "Don't spend the money to get [the consultant] out here. You get more right here [in the network] than doing that." Clearly, grazers recognized that they must adapt the "New Zealand style" to fit the very different climate, cattle, pasture species, and people that characterize local agroecosystems in Wisconsin; that is, they relied heavily on their own local knowledge.

Ocooch grazers also relied on publications and often recommended these to one another. Many members reported learning the principles of the technique not only from André Voisin's *Grass Productivity* but also from *Greener Pastures on Your Side of the Fence* by Bill Murphy (1991). *Greener Pastures* appealed to grazers because it was written in a clearer and more contemporary style than Voisin's classic work from which Murphy drew the core principles. Referring to the color of the book's cover, some Ocooch Network members indicated the guidance *Greener Pastures* gave them by calling it their "green bible." In this popular book, Murphy, who himself grew up on a Wisconsin dairy farm and who now farms and works at a sustainable agriculture program at the University of Vermont, provides useful information for grazers. "If I ever get into trouble," said John Hills, "I go back to that [book] because it's got everything in there that you could think about grazing, and it is pretty up to date." Grazers also reported reading and being influenced by other publications. A Wisconsin weekly agricultural newspaper, *Agri-view*, had published numerous articles about grazing in general and about different grazers' experiences in particular; the paper has thereby played a significant role in promoting and monitoring the development of grazing in the state. In fact, many of the network members remarked that articles in *Agri-View* were what first inspired them to switch to rotational grazing. Other important periodicals included the *Stockman Grass Farmer,* a publication based in Jackson, Mississippi. Before ceasing publication in 1995, the *New Farm Magazine* from Rodale Press routinely included articles on rotational grazing and on farmers practicing it from around the country, often featuring farmers from Wisconsin. Less frequently mentioned was *Holistic Resource Management* by Allan Savory (1988), which incorporates Voisin's grazing management as a key tool in a holistic, decision-making model. Short grazing fact sheets, videos, guides written by farmers such as Jim Brown, and network reports were also available and frequently distributed at network events. While recognizing the utility of these publications and other media, grazers explained that they also perceived limits to these sources. As Vicki Braun said: "Once you start making the commitment to do it, then you start doing the reading to find out about it. . . . You can gain through books, but it isn't the same as experience."

When network members referred to learning from experience, they meant not only their own personal experience but also the experiences of other farmers practicing the technique. One vehicle by which members of the Ocooch Grazers learned from other farmers was the Wisconsin Grazing Conference, which became an annual opportunity for experienced and novice grazers to exchange information for two or three days. Originally sponsored by the Southwest Wisconsin Farmers Research Network in 1992, responsibility for this statewide conference was transferred in 1993 to Grass Works, Inc., an organization whose board of directors included grazers from around the state, including Ocooch coordinator Mike Cannell. All but one of those network members I interviewed had attended the conference at least once, and several members made presentations over the years. According to one observer, "This conference is almost unique because it isn't bankrolled by some big organization with people more comfortable in suits than boots with manure on 'em." Indeed, the grazing conference, which was organized and attended primarily by farmers, attracted approximately five hundred participants in 1993 and more than six hundred in 1994 and 1995. As in previous years, farmers dominated the presentations made in 1995: thirty grass farmers (seven of whom were women), twelve people associated with universities (three of whom were grazers), a veterinarian, and an editor of a grass-farming publication. In the process of organizing these events, Grass Works developed into a loose federation of networks by distributing lists of existing groups and facilitating the formation of new ones.

As a result, the event was more than the presentations; it also became an opportunity for the wider grazing community to gather annually and monitor progress in their own development of practices and ideas.

This annual conference was, however, a rare opportunity for farmer-to-farmer exchange, occurring only once a year. Much more frequent was the less formal setting of learning from one's peers living nearby. Vicki Braun described how her family learned to practice rotational grazing: "I can say with confidence that we relied heavily upon other farmers. . . . It's been primarily individual farmers plus the grazing network where we've learned the most." Indeed, among the grazers I interviewed, learning from their peers who lived nearby was the primary source of information that they reported using. This knowledge exchange is the focus of the next section.

In asserting the validity of their own personal, experiential knowledge and the value of learning from others who also have experience with the technique, conspicuously absent from the list of knowledge sources mentioned by Ocooch Grazers was the land-grant university and other government agencies. Michael Hall summed up what seemed to be the general consensus among network members: "Within the grazing network, I think especially towards the university, there's a negative view, and that's because we feel that what they've been teaching for forty years is wrong and that their ability or willingness to study the grazing has been slow." John Hills held a similar view although he noticed a change at the university:

When I started, the university had very little to say about it, but now it seems like they're getting more and more into it. . . . I think the biggest problem with the university though is they're funded by companies that want them to test their products out, and this is not a company system. There's no product here being pushed, so there's no money involved. . . . And if there's no money, how are they going to test anything? That's why the networks have to do this. They're basically doing the work the universities normally do, but it's like the farmers are doing the testing. So I don't think the university is quite as involved in this, but they seem to be a little more and more into it now. I don't know where the money is coming from now to test this out. Maybe it's just the simple fact there's so many farmers that are getting into it, they have to adapt to it.

Likewise, Vicki Braun reported: "We haven't learned anything from, like university extension or university farm instructors, those people that you would typically go to to find information. But they are coming around now, and they are promoting grazing and actually teaching something about it. But they're doing it after we had learned it already."

Clearly, members of the Ocooch Grazers perceived a change in the university's willingness to research and disseminate information on dairy grazing. Indeed, the increased popularity of the technique among farmers and the accompanying publicity about the paucity of information available from the university (e.g., McNair 1992c) prompted new research at the University of Wisconsin. Most notably, the Center for Integrated Agricultural Systems brought grass farmers, extension agents, and scientists together for a research group on grazing-based dairy systems (Gale-Sinex 1994). From the perspective of network members, however, those researchers and extension agents who were willing to consider rotational grazing seriously were the exception rather than the rule. As Tom McMahon put it:

I think there's lots of people in the university that refuse to recognize the value of rotational grazing because they're threatened by the fact that they've spent all these years telling these people how to do all this. If they go on out and tell them all they're wrong, it's going to raise some hell. I mean there's a clique down at the university that's pretty antigrazing and cannot justify why they're antigrazing. And there's a few people in the dairy department that are beginning to change their view and look at grazing. There's a few believers.

Members of the Ocooch Network were willing to learn from those few researchers who were interested in rotational grazing, as they did by inviting a plant breeder from the University of Wisconsin to answer questions at their 1994 winter meeting. At that meeting, this plant breeder admitted that when some farmers asked him several years earlier to participate in an on-farm research project on grazing, he agreed reluctantly, "kicking and screaming." But in the process he slowly became, as he described it, "a believer" in rotational grazing.

While network members seemed receptive to learning from those agricultural scientists and extension agents who had something to offer them, their receptivity was tempered by their assertion that farmer-generated knowledge of grazing was considerably more advanced than that of the scientists. For instance, when reporting on the results of a research project, Joel McNair (1994) criticized the researchers for examining questions that grazers already knew the answers to or that were deemed irrelevant. McNair (1994:3) suggested that scientists should "seek more advice from working graziers. When it comes to grazing research, these are the people who've earned tenure." Ocooch Network member Glenn Scoville agreed with this message and expressed a feeling that seemed to be held by others in the group: "I've read their stuff, but generally the stuff that I've read is stuff that we did last year and the year before. . . . Networks are ahead of the university . . . by maybe a couple of years. Maybe even further."

What we see, then, is that the *principles* of rotational grazing are not indigenous to Wisconsin; they were conceived in France and were further developed by farmers and researchers in other countries. And the principles can be learned from publications, consultants, and potentially the university. But according to members of the Ocooch Grazers Network, the *practice* of grazing depends on the development of local, experiential knowledge. Mike Cannell was fond of comparing this knowledge to learning to ride a bicycle: "People can tell you how, but that doesn't mean you can do it. . . . You learn by doing." Referring to the fundamentally site-specific character of agriculture, Joanie Brown explained that "each farm is different"; therefore, network members constantly reiterated the need to adapt grazing principles to the unique conditions of particular farms. As Jim Brown said, "Most grazers don't adopt an idea, they adapt it to fit their situation." And that adaptation requires local observation and experience. Not surprisingly, then, Ocooch grazers came to understand that they themselves produced the local, experiential knowledge that must be the foundation of grass farming in the region.

Exchanging Local Technical Knowledge

At a day-long workshop in February 1995, interested farmers gathered in a small community hall to learn about what is involved in making a transition to intensive rotational grazing. As part of the program, several members of the Ocooch Grazers Network spoke about their experiences with the technique and answered questions on topics ranging from fencing systems to lifestyle changes. While the discussion of these and other subjects resembled many of the farmer-to-farmer exchanges that I observed during the course of this study, the event also provided a rare opportunity to hear some members talk publicly about the importance of farmer networking. Michael Hall articulated his perception of the group:

The main purpose is to obtain information and share information. There is a pasture walk once a month during the grazing season. We go to the host farm to see what's growing there, what's not growing there, whatever they want to show us. . . . I like to go see what ideas people have come up with. There are always new creative ideas. You can see what others are doing to cut costs. Farmers are experts at coming up with cheap, innovative ways to do things. And there are always new circumstances, not necessarily problems, but different things you have to handle. It is useful to talk to people about how to handle the wet spell or how to handle the dry spell. . . . There are always new questions. Chances are someone else has answers [and you] learn from someone who has had firsthand experience. . . . Most people are willing to show mistakes, and they're an expert at it. . . . It is time really well spent. And grazing is more management intensive, so the biggest thing is to talk to other people and try and avoid making mistakes.

Michael captured many of the central features of the Ocooch Grazers Network that will be discussed in this section. Attending pasture walks, listening to presentations at winter meetings, and reading network mailings provided clear evidence that the primary function of the Ocooch Grazers Network was to create opportunities for dairy grazers to learn from one another and sometimes from invited guests. Active members of the network constantly shared with each other their technical or substantive knowledge about how to graze. Given the tremendous emphasis the network placed on exchange of practical knowledge, a brief look at the scope of the technical issues covered may be useful before turning to a discussion of the knowledge-exchange process itself.

Technical Topics

Of the interactions I recorded during this study that constituted exchange of practical knowledge between members, roughly half dealt with how to develop and improve the pasture. My analysis of field notes showed that these fell into three subcategories: the general principles of grass management, the physical arrangement of the pasture, and the ecology of the pasture. By sharing technical knowledge about the principles of grass management, I mean such things as farmers describing the "tricks" they used to "measure" whether a paddock was ready to be grazed, like knowing the size of one's hand and using that to determine quickly the height of pasture plants. Or they developed and repeated certain idioms such as "When the grass grows fast, move the cows fast" to describe how to prevent the pasture from maturing too rapidly during the spring flush and thus to avoid the problem of the grass "getting away from you."

Exchanging knowledge about the physical arrangement of the pasturing system often captured the attention of members because a grazer's ability to follow the principles of grass management depended in part on how the pasture was physically laid out. Topics covered included how to subdivide the pasture by easily inserting lightweight fiberglass or plastic fence posts and stringing these posts with a brightly colored "polywire" (or "polytape") fed off a reel while walking; how to use this portable fencing technology to establish a temporary paddock of the proper size given the conditions at the time; and how to provide electricity to this fencing system, ideally using a high-quality "low-impedance" fencer that cattle "respect" and that will not ground out easily (a common complaint about the old electric fencing systems). Other physical elements of interest were establishing and maintaining lanes for moving cattle easily back and forth from pasture to barn, as well as different methods for providing water to cows on pasture. With respect to the physical layout of pasture systems, veteran grazers warned beginning grazers to "keep it simple," especially at first, and underscored the importance of customizing the system to conditions on particular farms. As they gained experience, grazers invariably found it necessary to change many of the pasture's physical elements over time; therefore, no one recommended setting up a permanent perimeter fence, building permanent lanes, or installing water systems in the paddocks during the first year. One of the things grazers said they appreciated most about this "flexible" technique was that it allowed for "refinements" and "modifications" as one's personal knowledge slowly developed.

The third and most recurring pasture-related topic of knowledge exchange in the network dealt with pasture ecology. In a mailing to members, Mike Cannell defined pasture ecology as "what grows, how does it grow, how do species interact, fertility, weeds, weed control, poisonous plants, species selection, [pasture] establishment, pasture development, [and] the impact of the hoof, clipping, [and] canopy control to favor certain species." Over the course of this study, all of these subjects and more came up repeatedly. For instance, I recorded numerous instances of grazers exchanging their observations on the advantages and disadvantages of particular species in a grazing system, as well as strategies for how to manage different species. Although pastures were often referred to as "grass," grazers thought pastures should ideally contain roughly 60 percent grasses and 40 percent legumes such as clover, which can contribute nitrogen to the pasture. But the question grazers grappled with was how to establish this desired ratio and how to maintain diversity in the pasture. Such diversity is key, they reasoned, because different species peak in production at different stages of the grazing season, and diversity affords protection from insects, diseases, and environmental changes such as drought, wet periods, and harsh winters. To establish this diversity, some members chose to rely on what they called the "seed bank" in the soil, emphasizing that grazers must learn to work with what is already there and use animal impact as a tool to encourage some species and discourage others. While not disagreeing that such a strategy would work, many others tried to speed up the process of species establishment by "frost seeding," that is, broadcasting seed directly into the pastures. Interestingly, as network members struggled to figure out what species worked best in their pastures, a redefinition of what constituted a weed emerged within the group. Alfalfa, "queen of the forages" in the confinement system, was disparaged by many grazers as a poor pasture plant. Conversely, quack grass, so long considered a weedy scourge of the corn row that it is illegal to sell seed, was recognized by grass farmers as a forage plant with superior nutritional quality and ability to withstand animal impact.

While about half of the knowledge-exchange interactions I observed dealt with the above pasture issues, another third fell into the category of milk production in grass-based dairying. Specifically this included meeting the needs of the livestock, seasonal milking, and low-cost milking parlors. With respect to meeting the needs of the livestock, network members exchanged their observations regarding the extent to which pasture forage provided sufficient nutrients and energy for lactating dairy cows and their calves. They debated what dairy farmers refer to as "balancing rations," sharing ideas on the degree to which pasture needed to be supplemented with dry matter, minerals, protein, or grain. Likewise, they discussed low-cost methods of feeding cattle during the winter months (e.g., silage piles). Other topics of interest were the grazing habits of the cow (e.g., which species cattle like to eat). Although network members seemed to agree that their animals had fewer health problems when on pasture than when in confinement, grazers occasionally discussed alternative approaches to managing those problems that did arise, such as feeding kelp for pinkeye in calves and using homeopathic remedies. Such alternatives seemed to be of particular interest to members who were certified organic producers or those considering such certification because organic production required stricter standards with respect to the use of drugs.

The second milk production subcategory, seasonal milking, refers to synchronizing lactation of the cows to the period of optimal grass growth in the area, roughly spring to early winter. In this way, grazers tried to take advantage of the low-cost forage provided by pasture and to reduce winter confinement feeding costs. What they referred to as "going seasonal" had potential benefits such as the possibility of a vacation when the cows do not have to be milked. But it proved to be an extremely difficult prospect, which became the focus of several pasture walks. Network members encountered problems in getting cows pregnant in a timely fashion so as to calve within the desired eight- to twelve-week period just before the Wisconsin grazing season begins, approximately March to May. Members exchanged strategies that they used to try to improve their skills in detecting cows in heat and to increase rates of conception. Simultaneously, they considered the role of cattle nutrition and genetics – particularly the genetics of the revered Holstein cow, which was bred for production, not reproduction – as the root of their dairy herd conception problems.

The third subcategory of milk production knowledge exchange focused on low-cost milking parlors, which were considered by many to be a "progression" in the development of the "whole system" of grass-based dairying. Grazers were interested in milking parlors as a way to improve the efficiency of the milking process over the more common mechanical milking units that are moved from cow to cow around a stanchion barn. In a parlor, by contrast, the cows move to the milking machines so that more cows can be milked in a given period of time. Such efficiency is particularly important to grazers because they want to maximize the number of cows being milked during the grass season. Milking parlors designed and built in the United States tend to be very expensive buildings intended for year-round use. By contrast, in keeping with their general commitment to low-cost production, grazers were interested in parlors used in Australia and New Zealand, which are much less expensive than most U.S. designs. When several network members built their own parlors in 1994, this topic dominated the discussion during two pasture walks at which members literally covered the "nuts and bolts" of designing, building, and using one.

Finally, though the vast majority of technical discussions in the Ocooch Grazers Network focused on the topics described above, a small percentage of the ideas exchanged among the members included specific suggestions on how to think about economic costs. This category included, for instance, discussions on how to calculate milk production costs when cows are on pasture in comparison to when cows are given stored feed during the winter months, for example, keeping machinery and building costs under the expenses for the winter because these assets are not used in the summer. In addition, several speakers at winter meetings emphasized the importance of trying to invest capital in "productive assets" such as land and cattle rather than "consumptive assets" such as machinery and buildings. Network members also discussed specific ways to save money, such as rolling polywire on inexpensive reels intended for electrical cord rather than buying the relatively expensive reels designed for polywire.

Socializing Local Knowledge

Ocooch Grazers exchanged knowledge about a range of topics related to the practice and development of grass-based dairying. An important characteristic of this information exchange was that members asked themselves what the appropriate questions were, and they looked to each other for answers. In so doing, they often drew on their own personal, local knowledge, which they shared with others who were similarly situated. This horizontal information exchange suggests, as Wainwright (1994) argued, that the common conceptualization of experiential knowledge as those things we know but cannot tell may have been overemphasized. Indeed, in this case, my observation of the network activities indicated, and in-depth interviews with the members confirmed, that local knowledge was repeatedly extended beyond the individual and meaningfully shared with others interested in intensive rotational grazing so as to increase members' chances of realizing their goals. Once network members articulated in the network setting an idea or observation derived from their own personal knowledge, that knowledge then became a social product available for use and interpretation by a community of knowers. In other words, local knowledge was *socialized* within the network.

Active members very much appreciated the knowledge-exchange process. Although grazers recognized that they must each individually gain experience and thus develop their own personal knowledge of grazing, members valued learning from others who had direct experience with using the technique. As Michael Hall expressed it, "Most practical information just comes from talking to other people that are grazing, primarily in the grazing network, because there aren't too many people right around this area here that are [practicing] intensive grazing." Similarly, John Hills said: "I've learned an awful lot. And there's almost no walk that I ever went to that I didn't learn something, if not from the farm itself, from people talking and sharing their ideas. I hate to miss a walk anymore." Or as Sally McMahon expressed it, "Every time you get information, not only from whatever speakers or whatever topic they're covering, but from the other farmers — what they've tried, and what's worked, and what hasn't worked, and throwing ideas back and forth."

This idea of learning from other farmers about what works emerged repeatedly during this study. For network members, one of the principal ways to learn what worked was trial and error, as Tom McMahon said simply: "You see things that happen, and if they work, you keep doing them. And if they don't work, you don't do it again." In turn, the network provided a valuable opportunity to share the results of that inquiry and to learn from the mistakes that others made. Michael Hall expressed it this way:

I've never gone to a pasture walk where I haven't picked up something I can use, whether it be an idea or information. . . . It's just so many ideas shared. And you can get any question you want to be brought up. Usually there's someone who's been at it longer that has asked that same question and that can answer it and share their mistakes, what they have learned, which is probably more important than finding out what did work, finding out what didn't work.

A willingness to share their mistakes indicated a certain humility among grazers, who readily admitted that there was much more to know. For example, one member remarked at a pasture walk: "One wonderful thing about this system is that it's very forgiving. We are all going to make mistakes, but we're learning." Of course, grazers want to make this learning process as painless as possible and avoid mistakes, as John Hills said: "I don't want to make mistakes, so why not learn from people that have. So that's another key thing there [in the network] is learning not only from your mistakes, but other people's mistakes." Similarly, Jim Brown reasoned that such learning was one of the network's purposes, as he stated at one pasture walk: "Let's not reinvent the wheel. . . . Let's not each one of us make the same mistakes."

Accordingly, at network events, members often shared with one another what they had learned from their mistakes. For example, during a June heat wave some forty people gathered at the farm of Glenn Scoville and his family for a pasture walk. Not long after the program began, Glenn explained that it was his third year grazing, and although "the pasture did great" in previous years, this was the first time the "pasture got away from me." He went on to describe to the crowd how he had fertilized the pasture in April, and when warm weather arrived in May, all the grass got very tall within a week. He continued to have problems later in the season, even after he had moved animals through the pasture rotation three times. The "mistake" of fertilizing in the spring was something he said he would not repeat, adding: "I guess everyone's in the same boat. We haven't been doing this long enough to know."

Network events provided an opportunity not only to hear what had and had not worked for others but also to see what others were doing. One could observe, for example, how well a particular clover had survived a harsh winter in the pasture on someone else's farm. Or one could inspect how a family converted their stanchion barn to a walk-through, flat barn parlor. As Michael Hall said: "We go to the host farm to see what's growing there, what's not growing there, whatever they want to show us. . . . I like to go see what ideas people have come up with." Glenn Scoville also endorsed the importance of creating opportunities for members to observe how something worked:

The reason I think these pasture walks really do help is because you can go see what

Members of the Ocooch Grazers Network listen to a field-day host.

a guy is doing right and see that it's working for him. Something you probably thought wouldn't work at all, but you can see it's working good. And then you can go home and say, "I can do that too." I've come up with some good information at some of them [walks], and then come home and it works for you just as well as it did for the other guy. You've maybe read about it and that, but you really didn't see it working.

For Tom McMahon, seeing what others were doing either confirmed some of his own decisions or raised questions in his mind:

I like to see what other people are doing. . . . I reinforce things that I do. Like I look at my pastures, and then I go look at somebody else's, and I say, "Theirs is shitty, mine is better. What I'm doing works better." Or if I go and see somebody else's pasture that looks better than mine, I say, "Well, what's he doing that I'm not doing?" Or "What's his soil like that mine isn't like?" Or "Why are his pastures better than mine?"

Thus one of the ways that knowledge was exchanged in the network was through demonstration and observation.

In addition to learning from others' mistakes and successes, network members also shared personal knowledge derived from their own creativity. During what might be called mini-demonstrations at pasture walks, members displayed gadgets or innovations that they had developed. At one pasture walk, for instance, several farmers said they used half-inch plastic pipe as a low-cost alternative to fiberglass fence posts, and one of the farmers demonstrated how he used a soil sampler to dig the hole into which he inserted these makeshift fence posts. Another farmer added that you can simply put a slit into the posts to run electric wire through and thereby avoid using a metal clip that could injure a cow. At pasture walks, Jim Brown was particularly fond of sharing the inventions he developed and the ideas he came across. Holding up a four-foot fluorescent light bulb, Jim explained at one walk that by attaching the bulb to an electric fence he could look out a window at night to see the light blinking when a charge was sent through it. Then he knew the fence was working, claiming it helped him "sleep easy" knowing the cows could not get out of the pasture.

During their exchanges, grazers raised a wide array of questions that had not been asked before and that would not be answered easily. Indeed, they did not have all the answers to the problems that emerged as they tried to create dairy systems based on this unconventional technique. The following are examples of recurring questions: What are the factors that make it difficult to breed cows so that they will calve during a certain time frame and so that the grazer can make best use of the grass season? Is it best to drag the pastures so as to spread manure patties around and thus minimize the creation of areas where the cattle refuse to graze, or does dragging pastures incur too much risk of spreading diseases and parasites? What is the best way to manage orchard grass that grows quickly and can outcompete other species in the pasture? Network members often had very different answers to the questions that were asked.

To illustrate, I attended many pasture walks and winter meetings at which network members articulated a range of views on the topic of whether pastures needed fertilizer in addition to the nutrients provided by grazing animals and legumes growing in the pasture. Some members thought that fertilization was advantageous, especially in periods of low moisture, while others felt that dry weather required a different management approach, namely, more time to rest the pasture between grazing rotations. One network member felt that fertilizer was absolutely necessary on his farm because of the need to maximize forage production on the relatively few acres he had available for pasture. Conversely, another member protested that in the book *Grass Productivity*, Voisin had been wrong in advocating the use of fertilizer on pastures. Still others thought fertilizer would increase pasture production, but if it were not needed, it would be smarter to save the money. In addition, some found they had trouble keeping up with what was already growing in their pastures; therefore, they figured that until they could manage what they had, it did not make sense for them to encourage more growth. Obviously, answering the fertilizer question was not an easy task.

At network events, members pooled their local knowledge in an attempt to answer the questions they raised. And they debated and disagreed with one another, both privately and publicly, about the answers to those questions. To handle the sometimes contradictory information that emerged in the process, network members reported that they had to sift through it and ultimately decide for themselves what worked on their own farms. For instance, John Hills explained:

You can experiment a little bit with it, and think a lot on it, because every farm has got a different situation – financially or the way it's laid out, or the cattle you have. . . . There's a lot of variations there, and you've got to adapt and make it work for your own situation. You've got to do a lot of thinking on it and see what will work and continue to listen to other people's experiences because that's what you need. You need input. . . . You look at your whole operation, look at the future, what will this do now and what will that change do in the future to my farm, and you can decide. Most of those things, if you think about them, they aren't that hard. Usually they're not really tough. Like I say, there are so many variations that you can adapt.

Other farmers agreed that the different viewpoints were not a serious problem for them. As Michael Hall put it: "There's always going to be little battles out there. It keeps it interesting."

Grazers' problem-definition process and the consequent attempt to pool their own observations and experiences suggested the possibility that through interaction over time, network members would not only exchange knowledge but perhaps would collectively create knowledge as well. The NRCS district conservationist, who came to many network events, observed about network members' competing knowledge claims:

There are contradictory things happening, but the grazing networks are the best solution to that problem. If you get three or four people together, and one person says, "Boy I did this and it worked" [and] if you get three other people that say, "Well, I tried that too and it doesn't work," just by the law of averages, the people who have tried the system communicate with each other, and you develop knowledge of the systems that actually do work.

As grazers gained confidence in their own capacities to generate understandings they did not previously have, they often referred to the future when they would answer the questions that arose. For example, a veteran grazer, who spoke at one of Ocooch's winter meetings, shared his belief that "we don't have well-developed knowledge about seasonal, grass-based milking, but once we figure it out, there is no way that bGH [bovine growth hormone] or California can put us out of business." This reference to grazers as a collective "we" that would "figure it out" was also articulated by a grazer and journalist who gave a rousing speech at the Tri-County Rendezvous, a winter meeting Ocooch cosponsored with two other groups: "There will be a renaissance in southwest Wisconsin. You'll be the envy of the ag world, perhaps the best grazing area in the entire country. We'll have great grass when we figure it out." He went on to compare grazers' skill development to that of his infant son, who would eventually learn to walk and talk: "Think about where your farm is going to be in twenty years. Right now we have a bunch of two-year-olds and a few three-year-olds, but eventually we'll learn how to do it right." Thus network sharing offered the possibility of producing new knowledge that no individual possessed alone. In this sense, the network's function was not only exchange of knowledge (i.e., socializing local knowledge) but sometimes knowledge creation (i.e., socializing knowledge production) as well.

The network's knowledge-exchange process had an egalitarian character. The coordinators, Mike and Jim, seemed to be negotiating a perceived boundary between wanting to make information of potential interest available to members and yet at the same time not wanting to appear to be "promoting" any particular certain method or technology. It was unclear to me exactly why this issue emerged, but at several pasture walks Mike and Jim made a point of saying that they were providing an article about a particular type of clover or giving out complimentary copies of a grazing publication, for example, but that they were not necessarily "endorsing" these practices. At one pasture walk Jim showed some fly traps he had found useful in the barn and explained how people could get them. He thought the fly traps were better "than spending a lot of money on poison," quickly adding, "I'm not selling any of this stuff, it's just that we could do with a lot less pollution."

Mike's approach to sharing information was even more cautious than Jim's. Although Mike had been grazing longer than most other members in the network and had demonstrated considerable grazing skill, he was nonetheless very reluctant to take on the role of "expert." In discussions with me he referred to this several times, and in the interview he articulated his view on the distinction between making specific recommendations and sharing experiences:

It is satisfying to me to know that I can help someone understand [intensive rotational grazing] a little bit better. . . . But I do not want to get in a situation where I think I know everything, where I'm going to tell you how you have to do this. That makes me an extension man, see, if I'm going to come out and tell you how to farm. I can tell you about my experience and what we've gotten out of it that was beneficial, and you can decide whether or not you want to do it. I'm not going to try and talk you into it. I'm just going to share the story, and you can decide. So to me there's a distinct difference between the two ways of looking at it. . . . It is not a situation where you have a teacher at the front of the room, and I'm going to dispense with this knowledge onto these students, and they're going to absorb it, and I'm going to see if they do with a pop quiz. It's a situation where a group of equal people come together, and we share stories, and we share experiences, and we share failures, and we share weaknesses. It's a situation where we can share all of these experiences and then use those experiences to build on our own knowledge base and make that knowledge that we've gained conform in some way to what we have in reality out here on our own farms.

By emphasizing sharing "the story" rather than imparting knowledge, Mike's approach to knowledge exchange is strikingly similar to that of critics of the dominant style of education, such as Paulo Freire (1970) and other advocates of participatory research. Mike's rejection of the unidirectional flow of information from the "extension man" who dispenses knowledge to the practitioner may be compared with Freire's (1970:58) description of the "banking style of education" in which "the scope of action allowed to students extends only as far as receiving, filling, and storing the deposits." Instead, Mike saw all network members as "equals" who shared their experiences. As a result, Mike consciously resisted the tendency to develop certain individuals into what he called "grazing stars" because he was well aware that within the Wisconsin grazing community in general a few grazers had gained the reputation of being "masters" (see also Rittmann 1994). By rejecting the role of expert, Mike's view corresponded to a key feature of new social movements that Wainwright (1994) observed; that is, it reflected a distinctive approach to knowledge that simultaneously asserted the validity of experiential knowledge and refused to claim an all-knowing expertise.

Within the network, differences in physical location were often recognized (e.g., how soil type might affect the way one answered the "fertilizer question''); however, differences in social location were acknowledged much less frequently. Social diversity inevitably exists in a given locality (Feldman and Welsh 1995), and in this case, gender seemed to account for some distinct differences among members' participation and interactions in the network. Men tended to outnumber women at pasture walks and other network events, usually by a ratio of about four to one. One possible reason for this gender imbalance was that farm women often held off-farm jobs to supplement the farm income, whereas none of the five women who came regularly to network events were employed off the farm.

Given the fewer number of women who came to network events, it may not be surprising that ''usually men do the talking,'' as Sally McMahon put it and as I observed. Of course, the fact that men were in the majority may not be the only explanation. Some women felt that they participated even less than their numbers warranted. Asked whether she saw any differences in the way that women participated in the network compared with men, one woman replied:

Us farm wives do a lot of the work. We do a lot of putting the cattle out to pasture, but we're not involved in the network itself. We're not up there talking and telling other people. Now, I wouldn't mind doing it, getting up there and telling them how I do it. I'd be glad to, but nobody ever asks me. I think it's that they are just overpowering. We stand by, as usual, just sit by and let them talk. . . . I think women should be more involved. I think women should get up there. I think that women should be allowed to participate more. Like with [my husband] and I, he gets up there and gives a big talk on what he's doing, but yet I'm involved too. I guess he speaks for both of us, but I don't know. Really we both own everything, and we both do everything, and it's both our ideas. . . . I think it's something that they should have a women's point of view. The women should get up there and talk about what they've learned and what they think should happen, instead of just having it all the men. . . . I do all the feeding, I've driven the tractor, I've baled hay, helped clean the barns, clean calf pens. I get in on everything, but when it comes to getting the credit, I guess he gets all the credit. That's the way it is.

Another difference in men's and women's participation in the network seemed to parallel some common divisions of labor in family dairying. On family farms, as elsewhere, men and women tend to do different kinds of work. Charlotte Cannell indicated as much:

While women will move the fence, it's normally the husband or the man's job. I don't know why, other than the women are probably feeding the calves. So we let the men take over talking about the grasses, but not always. . . . We're listening and learning [in the network], but it seems the application of it is probably, I feel, more men-oriented. Although the women there, we're sharing our stories, we probably wouldn't get up and talk in front of the group. We might talk about forgetting to open a gate and how the cows react to that, but most women aren't discussing the water or what's going to be planted.

Indeed, women did not actively participate in network discussions on seed selection or other technical topics as much as men did.

Women were sometimes very active, however, when sharing with one another their own experiential knowledge regarding those production-related activities for which they did take primary responsibility, such as tending young calves. On most Wisconsin dairy farms, calves are weaned from their mothers soon after birth, raised individually in separate hutches, and fed milk, initially from a bottle and later from a bucket. This work, frequently done by women, is time-consuming and expensive, especially when many calves are born around the same time in a seasonal system. At a pasture walk in September 1993, an alternative known as the "calf-rearing barrel," popular in New Zealand, was demonstrated and discussed by the members. In that system, groups of calves get milk by sucking on nipples attached around the circumference of a plastic drum with tubes that carry the milk up from the bottom of the barrel. Several women who fed calves raised skeptical questions about the method (e.g., whether calves would suck on one another if allowed to be grouped together) and expressed their reluctance to switch from their current methods.

Although Charlotte Cannell was one of those who was hesitant to try the calf-rearing barrel, she went to a nearby farm where another woman was using it to see how she did it: "I saw her putting her fingers in the calf's mouth and then on the nipple [to get it started]. And those [calves] that knew what to do were up there drinking vigorously, and one was just being introduced. Then I felt comfortable with it. So the seeing had a lot to do with it, for comfort, to know it can be done." When she did try the barrel, Charlotte found she really liked the technique, and at a pasture walk the following spring, I observed her sharing her experiences with another woman member who also had primary responsibility for calf rearing but who had not yet tried the calf-rearing barrel. Charlotte described, for example, how the calves seemed healthier and how it took "attention" during the first few days to be sure each calf was getting what it needed. Tracing the knowledge-exchange process in this way reminds us that the production and exchange of local knowledge in this network may often reflect the gender division of labor on these farms.

In sum, in the Ocooch Network, members traded ideas on a range of technical topics related to grass-based dairying. The source of much of what was exchanged was their own local knowledge, exemplifying Wainwright's (1994) suggestion that personal, practical knowledge can be extended beyond the individual and become a social product.

Identifying as a Grass Farmer

In my observations of the activities of the Ocooch Grazers Network, I found that members' exchange of practical knowledge about *how to graze* was complemented by their exchange of ideologies, that is, values, ideas, and beliefs, about *why to graze*. Eyerman and Jamison (1991) argued that social movements tend to change how participants interpret the world. In the network, one manifestation of such reinterpretation was that many grazers no longer thought of themselves as *dairy* farmers, but rather as *grass* farmers. Some grazers sported hats and T-shirts identifying their owner as a "Grass Farmer." Jim Brown explained: "I market grass. I sell it in the form of milk, but my raw material is grass." One grazer was fond of remarking that "this isn't about grass, it's the way you think about things." Among grazers, this reinterpretation of who they are and why they farm was intimately related to a retooling of their operations and a reorganizing of their relations with one another. In so doing, grazers implicitly rejected the ideology and technical trajectory of conventional dairying.

In the discussion that follows, I explore some of the most prominent dimensions of this "dairy heresy" (McNair 1992a) that I observed during the course of interactions among network members and that were articulated by individual members in interviews. Many of the perceived advantages of grass-based dairying over confinement feeding that I explore here have also been reported in academic and movement literature (e.g., Gage and Smith 1989; Krcil and Gralla 1995; Liebhardt 1993; Shirley 1993; Wisconsin Rural Development Center 1995). In describing the perceptions of members of the Ocooch Grazers Network, I want to make it clear that not all members held all of the views discussed here. Accordingly, I give some indication of how members differed in their adherence to these alternative values. Yet by providing an opportunity for people to share their different views, the exchange of ideas about why to graze revealed that networks themselves can lead to transformation in how some participants think about farming.

Productionism, Profitability, and Rural Community

One profound heresy in rotational grazing is that it means adopting a very different way of thinking about production and profitability than that associated with the sociotechnical system of conventional dairying. Maximizing milk production per cow throughout the year is the primary focus of conventional dairying in Wisconsin and elsewhere in the United States (Fales 1994). To maximize production, private and public researchers maintain a steady flow of innovations for the technological treadmill, bovine growth hormone being a recent example (Campbell 1993). To counter the challenges from expanding industrial dairies in California and Texas, increasing costs, and falling milk prices, Wisconsin's dairy farmers have been forced to produce more milk by intensifying their own labor (e.g., by milking three times a day), by running faster on the technological treadmill, and by expanding herd and farm size. The only alternative often seems to be to get out of farming entirely. Between 1988 and 1993, Wisconsin lost over six thousand dairy farms and experienced an increase of five hundred farms with more than one hundred cows (Galloway 1995:1E).

If there was one thing that network members did agree on in answering the question of why they practiced intensive rotational grazing, it was a rejection of what Buttel (1993a:7) referred to as the "productionist ideology" in agriculture, that is, "the doctrine that increased production is intrinsically socially desirable and that all parties benefit from increased output." Such a "mentality leads us to a false sense of prosperity," according to Jim Brown, because "those of us who have been listening to the swivel chair experts . . . can't figure out why we can't work hard enough or long enough or invest in enough equipment and fertilizers to make all our dreams come true." To illustrate the situation dairy farmers confront, Mike Cannell conjured up the following image and repeated it in a variety of settings:

> I think that everybody is under stress, specifically in the dairy industry, when we see a thousand farms a year going out of business [in Wisconsin]. That also means that there are also four or five or six thousand farms that are close to going out of business. I talk about all of us dairy farmers being lined up at the edge of a cliff, and we keep dropping off the edge a thousand a year. . . . But if I can find ways of change, in management or methodology, that allow everyone to step back from that cliff, that's what I'm trying to do. I think that grass dairying, seasonal milking is one good way of doing that. . . . I am looking for technologies that are really management based and not capital based.

Before switching to grazing, some network members had indeed come dangerously close to dropping off the edge of the metaphorical cliff. John Hills echoed many other network members whose operations were precarious: "I can make more money from grazing than I could before, so that's why I do it. . . . I was either going to quit farming or go to grazing. It was that bad."

For grass farmers, opening space to move back from the cliff's edge meant rejecting the focus that most dairy producers place on using large amounts of capital and off-farm resources to maximize production per cow throughout the entire year.

Glenn Scoville was one of the network members who had previously listened to the "experts" and followed the productionist ideology closely. Now in his late fifties, Glenn and his wife, Dolores, first started dairy farming in 1967. Their story is similar to those of many other farmers who focused on increasing their milk production considerably by expanding their operations in the 1970s. The Scovilles built a free-stall barn in 1970, only to add an expensive milking parlor four years later when they increased the number of cows they were milking. Deciding to rely completely on stored forage, they bought two Harvestores – top-of-the-line, glass-lined metal silos for storage of high-moisture crops – that "cost a lot of money" and stand as visible symbols of investment. By 1980 the Scovilles had purchased another farm nearby. Like so many other farmers who had adopted what Peggy Barlett (1993) referred to as an "ambitious management style" in the 1970s and 1980s, the Scovilles saw their net worth plummet when interest rates skyrocketed. As a result, Glenn said, "We had a pretty tough time in the 1980s. . . . Instead of paying off any debts, you was lucky if you just paid the interest for a while." The value of their second farm dropped dramatically, and, according to Glenn, "it's only worth about half today what it was when I bought it." Yet, unlike so many others who lost their farms, the Scovilles managed to make it through those crisis years by taking a more "cautious management style" (Barlett 1993) and with help from Dolores's off-farm job.

Looking back, Glenn felt he had been "pretty foolhardy. . . . Farming was really good to us in the seventies, and then we built a lot and got too brave, listening to all the farm magazines and university people. They predicted a lot of different things that didn't happen, especially [that] the price of milk would continue to go up. . . . Pretty near everybody that was *not* farming was encouraging that kind of expansion." One lesson he learned was "don't pay much attention to the forecasters." But another lesson was one repeated over and over again by Glenn and many others in the network: "Don't spend money, if you don't have it and if you don't need to." Therefore, adopting intensive rotational grazing in 1991 fit into the family's more general commitment to pursuing lower-cost production. Although Glenn cited a variety of different benefits from his family's adoption of rotational grazing, the principal advantage was as a cost-effective alternative to the productionist strategy he had so closely and so dangerously pursued earlier.

Whether network members had been "deep in debt" or not, they seemed to share a rejection of the belief that capital investment is necessary for profitability. Instead, grazers favored the idea of using on-farm resources and practices – cows, pasture, and the "grass eye" – to boost profitability through lower costs, even if this meant lower milk production per cow (although such reductions did not always occur). Their belief in the success of this approach was expressed in a logo on the T-shirt of a farmer who once came to an Ocooch pasture walk from his home located several hours away. The logo read: "I'm proud to be a grass farmer. Now if you will excuse me, I really must get to the bank," and it inspired cheers of agreement from others at the event. For some network members the conception of profitability extended beyond their own personal financial gain toward a general sense of maintaining the viability of the rural community. Specifically, grass-based dairying was viewed as advantageous for rural areas because improved profitability could help keep existing farmers on the land and because young farmers could get established without having to make large capital investments, which many hoped would reverse the current decline in the number of people entering farming.

In some instances, grazers' alternative conception of the route to profitability constituted a direct challenge to the agribusinesses that have benefited most from the introduction of new technologies over the last century. Network members, like Jim Brown, appreciated that they did not "have to go to the bank or machinery dealer to try an idea out." Unlike magazines that present "what the companies want you to think," John Hills felt that what he learned in the network helped him recognize that he might not have to buy those things. "Automation," said Glenn Scoville, "makes you a slave. . . . All you do is fix the things . . . and they always break down when the weather is the worst, the coldest, and the nastiest. . . . The farm papers and the university would say, 'Buy automation and do it cheaper.' Well, it works great when it's new for a little while, but then it don't stay new very long, and generally you find out that what you buy hasn't really improved your profit margin that much." The recognition that in grass farming you use your "eyes and brain rather than the power of your pocketbook" empowered grazers to act on the agrarian populist insight that, as one farmer unapologetically put it, "we don't owe agribusiness a living." Instead, grazers themselves produced most of the knowledge they needed. Discovering that they could learn from and teach one another, network members developed a sense of epistemic self-reliance (Hassanein and Kloppenburg 1995).

Pleasure, Work, and the Grass Bug

Another central tenet of the grazers' worldview was the value network members had come to place on the improved "quality of life" and better "lifestyle" grazing offered them. In fact, many found their work to be "fun" and "enjoyable." One farmer began the pasture walk at his farm by explaining that he had fed his cows in confinement for twenty-two years: "I got more milk, but I also ended up with more work and lots of machinery to take care of too. I kept wondering what was going on." When he read several articles about rotational grazing and attended a couple of pasture walks, he decided to switch. The result, for a man who had farmed since the 1940s, was that "pasturing my cows makes farming fun again." Frequent references to the "fun" encountered when transitioning to intensive rotational grazing reflected a strong sense of excitement among grazers and made it clear why grass farmers talked about having the "grass bug" or "grass fever."

What made pasturing fun for these farmers? Not surprisingly, how the grass bug expressed itself depended on whom you asked. Unlike confinement feeding, "it isn't such a rat race," according to John Hills, "you can look around a little bit and enjoy life. Otherwise, you're just looking straight ahead and thinking we got to get this done, we have to get that done." Rich Braun also appreciated abandoning the frenetic pace: "Part of this whole thing is walking, and we enjoy it." Some of Glenn Scoville's fun came from what he could hear:

One of the most fun things to do after you've begun grazing is, especially at night, you turn your cows out, and then you go up to where the cows are in a pasture, and you can hear all of them cows chewing. If there's a hundred cows, you wouldn't believe the noise they make. They're just eating that grass. I mean it's almost unbelievable. They'll be spread out over that whole pasture just mowing that off there, that's how contented them cows are. Then pretty soon, one will lay down, and after a while they'll all be laying down there chewing their cuds. That's some of the enjoyment out of it too.

Whatever their reasoning, network members seemed to agree that certain aspects of grazing were fun.

When they compared intensive rotational grazing to confinement feeding, however, there was some disagreement on the amount and character of the actual work required. Most commonly, grazers claimed the new technique reduced their work of planting, harvesting, transporting, and feeding crops during the grass season, when cows harvest forage for themselves. This perception of the labor savings was so strong that apparently there were rumblings among conventional farmers that grazers lacked a strong "work ethic" and that they were "lazy," as several farmers reported. But for Joanie Brown, grazing "sure beats feeding with the wheelbarrow in the barn every day of the year." As Vicki Braun pointed out: "You are not using any machinery to harvest, so you don't have to start the tractor up and that type of thing. You just take the animals to where they need to harvest, and they do the harvesting, and you watch it being done. It's great."

A similar perception was the primary factor motivating at least one Ocooch member to adopt the technique. While standing on a ridge-top pasture at a farm finishing up its first season of grazing, network members asked the host farmer why he had decided to switch. Pointing over to his neighbor's farm, which had been in pasture for seven years at the time, he shared his story. As on so many other days, he had been out "green chopping," a process by which the farmer cuts just enough of the hay crop each day during the summer and hauls it directly to the cows. But one day, he watched his neighbor come out of the house, move the fence, and go back inside. Still in the field chopping forage, the host farmer found himself thinking about the difference in the amount of work each had to do to feed the cows on that same day. Observing this difference was the "number one reason" he switched. Network members also found that adopting intensive rotational grazing required them to handle manure much less frequently because cows spend time in the barn only during milking. And less use of machinery translated into fewer days spent fixing it when it inevitably broke down. To the extent that grazers managed to synchronize the breeding schedule of their cows for seasonal milking, the opportunity to "dry off the herd" and take vacations when the cows do not need to be milked was much appreciated. Indeed, for many network members, adopting rotational grazing appeared to reduce their labor requirements in these and other ways.

Others pointed out, however, that rotational grazing did not necessarily reduce their workload; rather, it required different kinds of work, especially initially. When first switching to rotational grazing, farmers had the additional work of installing pasturing systems, such as building perimeter fencing, constructing permanent lanes, and eventually putting in water to the paddocks. Before these systems were in place, grazers found that moving cattle and providing for their water needs was time-consuming. And a few grazers pointed out that some of their pastures were located a considerable distance from their barns, so time spent going out to the paddocks and bringing cows in for milking was an added labor need, sometimes requiring them to get up as much as an hour earlier in the morning.

Perhaps the biggest difference in the work required by rotational grazing related to the fact that, as Michael Hall put it, the technique is more "management intensive." Others similarly stressed that rotational grazing requires careful attention to both the cows and the pastures. For this reason, some grazers preferred the terms, *management-intensive grazing* or *intensive rotational grazing* over the more common *rotational grazing*. They argued that the word *intensive* more accurately captures the essence of the high levels of concentration and attention required to practice the technique successfully. Glenn Scoville observed that with grazing, "You have a different kind of work. . . . We have had more free time since we went this way. But some of the people say 'Well, you just turn them out to pasture and that's it.' I don't think there's any less management; in fact, maybe you have to be a little quicker on things, although it is something that is forgiving." For Jim Brown, who felt that "observation is the key to life," the work involved in rotational grazing meant "I don't get to throw a switch to feed my cows, I have to walk out in the pasture and observe, and think how much to give them today and tomorrow. And answer one thousand questions, like, 'Grandpa, what kind of bird is that?' " Obviously, matching pastures to the needs of livestock was materially important to members of the Ocooch Network, but catching the grass bug also had to do with the pleasure experienced in successfully meeting that challenge, perhaps especially with one's grandchild in tow.

Healthy, Happy Cows

If network members found pleasure in their work, many were also thoughtful about the ways that adoption of rotational grazing was easier on their livestock. In the course of network events, many members observed that their animals' health had improved when they practiced rotational grazing. For example, although Vicki Braun expressed concern about what she thought was an increased incidence of pinkeye in their young livestock when on pasture, she generally felt that grazing was "good for the cows physically because they get the exercise and because they get to be outside full-time during the nice time of the year. They get fresh forage which is by far higher in vitamins and nutrients than stored feed." As a result of such exercise and fresh forage, many grazers found that their animals had fewer health problems than when confined to muddy or concrete barnyards (notwithstanding the initial bloat problems experienced by the McMahons). Overall, most grazers found that improved animal health naturally translated into fewer visits from the veterinarian, which saved them money. And because cows stayed cleaner while on pasture, grazers found the cows were less prone to mastitis. This in turn enabled the grazer to reduce use of antibiotics and thus also reduce the need to "dump" the cow's milk as required when the animal has been treated.

While I had expected the network members I interviewed to cite these animal health reasons for why they grazed, I had not anticipated their expressions of appreciation and caring for the animals. Mike Cannell articulated the strongest pro-animal philosophy:

I thoroughly enjoy giving the cows an opportunity to be cows. I want the cows to be able to enjoy this cowness that is their very nature, and in a confinement operation they cannot. I mean they are just little milk-making factories. Confinement operations do not treat cows like cows. So that relates back to whatever respect I have for other species, and I have to say to you that philosophically I do not put the human species above any other species. I just don't. And I have said to you that I'm against using animals for research, and I am. I don't think we're any better or any worse than the termites or the baboons or the cows. We just happen to have more brains and more ability to manipulate the natural environment in which we live. But it doesn't mean that we should be immoral or unethical or cruel. We have the ability to be that way if we want, but we also have the responsibility to not be that way. And so I feel that the cows out on grass is a natural thing to do.

While Mike identified with the rights of other animals, most network members simply stated that one of the reasons why they grazed was that their cows seemed "contented" on pasture. For instance, Michael Hall expressed his "really strong love" for Jersey cattle and said grazing gave him the opportunity to watch them: "I just love watching cows eat grass. They're out there happy and eating. Especially when the sun is shining, I mean I don't like standing out there in the rain, but I just enjoy being out there."

Conservation and the Environment

Absent the need to plow, plant, till, harvest, and store traditional feed crops, rotational grazing greatly reduces the need for and thus the environmental impacts of specialized machinery, petroleum products, and agricultural chemicals (Murphy and Kunkel 1993). Curiously, environmental protection is one of the most oft-recognized goals of the sustainable agriculture movement, but members of the Ocooch Grazers varied in the extent to which they articulated an environmental ethic. Some members were clearly motivated by the real and perceived environmental benefits of grass farming, while others recognized those benefits but did not necessarily think they were of utmost importance.

During observations of network events, the most frequent reference to environmental stewardship that I recorded dealt with a perceived reduction in the soil erosion associated with annual or biennial cultivation of a single field crop. Perhaps this is not surprising. Since the 1930s, when ecologist Aldo Leopold participated in a soil conservation project in this very area (Meine 1987), soil erosion has been recognized as a major concern in the Driftless Area of Wisconsin, which is characterized by steep hills and narrow valleys. For instance, John Hills knew the steep land near his barn "needed a little more conservation," and he thought grazing would be the best way to achieve that: "If you never have to plow it up, you can get a thick matt on the ground where you don't have the soil exposed." Similarly, referring to the steep slopes that characterize much of his ridge-top farm on the bluffs above the Mississippi River, the farmer who switched to grazing after twenty-two years of confinement feeding told the group, "I feel like I've been working up land that should only be in pasture." After making the transition, grazers observed the improvements. For example, while walking through his pastures in 1994, one farmer told me this interesting story. One day, when it started raining hard, he was out in the pastures, and his brother was over by one of their family's cornfields. His brother told him later that he could see the runoff from the cornfield as soon as it began to rain. But the farmer relating the story to me had noticed that there was no runoff coming from the pastures; instead the water percolated into the soil. With their own eyes, they saw the benefit of rotational grazing over corn cropping in its reduction in soil erosion. In turn, they shared the story with others.

In addition to soil conservation benefits, taking land out of monocultures of corn and alfalfa and replacing them with biologically diverse pastures reduces the need for agrichemicals and petroleum (Murphy and Kunkel 1993). But the degree to which members of the Ocooch Grazers Network saw decreased chemical use as an advantage varied considerably. Some network members have always been low-input or certified organic farmers, and several marketed their milk through a farmer-controlled organic dairy cooperative. For Rich Braun, organic farming continued "a family tradition" on their farm because the uncle from whom he took over the farm rejected the "chemical revolution" in the 1950s. When Vicki and Rich began farming there, they never used chemicals because "organic fits with our concept of stewardship of the land and the way we want to leave it for future generations to come." Other farmers did not have such a long tradition of avoiding agricultural chemicals, but a few remarked that adopting grazing made them subsequently question the need for agricultural chemicals, as they began thinking about alternative means to profitability, keeping farmers on the land, and using their skills to create healthy, diverse pastures. For instance, John Hills put it this way:

I think grazing is one way to stay sustainable. That's why the last couple years "sustainable" is becoming a big word in my vocabulary. Now I'm getting more and more where I don't even want to use chemicals, so I'm almost leaning towards organic, but I don't know that's kind of maybe down the road a ways. But anything I can do to avoid the chemicals, I do. Even the drugs in my cows. I used to just get the old syringe out and shoot them all up, but if there's a way to avoid it, I do. It seems like they're healthier. I think we're going to be shifting away from chemicals more and more. Either that or it's going to be one farmer running the whole world with chemicals, and I don't think that will be a very fun world. . . . If you're really thinking about preserving the environment, you have to have people to do it. You can't just say, "Hey, I can farm five thousand acres." . . . It takes people to take care of the land.

Not all members expressed the same appreciation for reducing chemical usage. For example, when reporting why he practiced rotational grazing, Michael Hall mentioned profitability, forage quality, milk production, less work fixing machinery, and animal health. Environmental concerns in general and reduction of agricultural chemical use in particular were absent from his discussion. When his use of agricultural chemicals came up in a different context, Michael explained that he raised corn for silage to feed his cows in the winter. As a farmer who carried a lot of debt, finances – not the environment or human health concerns – were the sole basis for his decision making: "Not that I'm going to spend a lot of money [on chemicals], but if I can spend something and it's going to pay a good return, well, I'll do it regardless of anything else probably."

While all network members shared a concern for the economic viability of their farms, some also associated an environmental ethic with their approach to milk production, or what Jim Brown referred to as "natural farming." Asked why he practiced grazing, Tom McMahon said:

It sounds stupid, but I feel like you have an obligation to the land. It's kind of a Zen or Native American view of the land. You only own the land for a short period of time, and while you own it, I think you have a responsibility to maintain it so that it's still productive and usable and not poisoned for the generations in the future whether they're people in your family or not. I think rotational grazing fits into that view of the land better. And I think it puts the cows into a more natural system. . . . The predecessor of the cow, which nobody knows what the hell it was, probably grazed, like the buffalo did in this country, and moved on and basically rotational grazed as a natural way of existing. And they probably calved in the spring like most other animals, and I mean it's just a more natural system. I think that it's less costly both economically and environmentally to try and have the system be in sync with nature. It doesn't make sense to calve cows when it's twenty below zero in January. It just isn't natural. . . . I think that you're better off not fighting nature. You can try to get her to work with you.

At the same time, however, Tom also recognized that "you can do a lot of things that are very environmentally damaging as rotational grazers." Mentioning that his cattle have to cross through a creek to reach their pasture, he explained: "If you look at the mud hole out in front of this place, I mean there's proof positive that rotational grazing doesn't necessarily have to be environmentally sound. It depends. I mean our particular problem is that we have so many cattle currently on this farm that there's nothing we can do about it. And that's why we bought another farm."

Grazing Is Believing

On the whole, the wide variety of real and potential benefits of dairy grazing that network members identified during the course of this study reflected a proactive, empowered outlook that stood in sharp contrast to the attitudes of many conventional producers. Patrick Mooney and Theo Majka (1995) have shown that agrarian thought and activism have long harbored a sense of farmers as pawns, whether it be of monopolies, industrial society, government programs, or bankers. However justified those concerns might be, at gatherings of grazers I observed little talk of grievances. Rather, there was a noticeably positive tone to the proceedings which at times verged on the messianic.

The practically unqualified belief in the benefits of grass-based dairying prompted me to ask network members if they had encountered any drawbacks from adopting the technique. It is hardly an understatement to point out that they had comparatively little to say. Several members simply could not think of *any* downsides to adoption of the technique (and even asked me if I knew of any). Some challenges have already been mentioned such as preventing bloat in cattle, spending more time to move cattle to and from pastures that are far away from the milking barn, and getting cattle bred for seasonal milking. The other issues raised were "pretty minor" and were not "real big drawbacks," according to network members. One farmer speculated that if there were not open land relatively close to the barn, it would be difficult to adopt rotational grazing. Weather-related issues were mentioned by three grazers, such as the observations that during extremely hot weather cattle grazed very little and that when the weather changed, daily milk production fluctuated. Finally, like many other grazers who turned a hurdle into a benefit, Vicki Braun mentioned: "One drawback is we have to climb a very steep hill [to get to our pastures], and if you aren't good at walking, that would be a real drawback. But for us it has ended up being a benefit too because it has given us such good exercise."

The local NRCS district conservationist also noticed a lack of criticism among grazers:

You go to the grazing meetings and you go to the [annual] grazing conference in Stevens Point, it's like a bunch of Amway salespeople. They are all so high about their product, you don't really get a full story. But out of all the grazing people I talk to, you just don't hear very many bad things about the system. . . . There just seem to be so many positive things that I haven't heard any negative things. Now maybe it's because the people that are doing it have bought into it and don't want to admit that it's not working. But there's not that many people that I know of that have done it and gotten away from it.

Tom McMahon explained this phenomenon: "Rotational grazing is really a religion. It's like people that quit smoking become antismoking activists generally. People that begin to rotational graze, it's like a light lights up over their head. It's like one of those 'ah-ha' moments. It really becomes almost a religious experience. I mean people go from one extreme to the total opposite and become believers. . . . And the reason is that rotational grazing works."

In sum, as these dairy farmers began to identify as grass farmers, they simultaneously challenged many of the prevailing ideologies and assumptions that characterize conventional agriculture. In generating and exchanging local knowledge for sustainable agriculture, network members not only thought *for* themselves as they engaged in practical activity but thought *about* farming in new ways. In other words, identifying what works in a practical sense also means identifying what works in an ideological sense; that is, generation and exchange of knowledge is intimately related to how one sees the world.

Building a Supportive Community
In my observations of the Ocooch Grazers Network, it was clearly evident to me that the group functioned so that members could trade local knowledge about how to graze, and in the process they also exchanged beliefs

about why to graze. In the course of the interviews, however, network members made me aware of another dimension of the network's function that had been left unspoken at network events, namely social support.

By rejecting many of the technologies characteristic of confinement feeding, members of the Ocooch Grazers often found that they became disconnected from their immediate neighbors who continued to farm conventionally. This disconnection was so profound that some farmers reported feeling ridiculed, and they repeatedly used the word *crazy* to describe how they thought they were perceived in the rural community. "I'd go into the feed mill down there, and everybody would shut up," explained Jim Brown. "I had several people tell me that I am crazy at the beginning. They'd say: 'You're nuts. Why are you doing that? You can't get no milk out of them cows.'" Even some grazers admit they had the same initial impression. For instance, when John Hills first learned about Jim Brown's adoption of rotational grazing,

I thought he was crazy. . . . Everybody would have eight-foot-tall corn and three-foot-tall alfalfa. And he would have little five- or six-inch grass there, and I thought, what is this guy doing? The whole farm is ruined. And, lo and behold, that's what my neighbors are telling about me now, that I've ruined my farm. Little short stuff, but I get a lot more crops than they do. I get like seven or eight crops, and they get one or two crops. So, lo and behold, I'm crazy too.

Glenn Scoville tells a similar story: "At first I thought it was crazy, thinking you could graze as long as they say in Wisconsin. And my son, he thought it was even crazier yet. He wasn't at all for it. He thought it was nuts. He thought *I* was nuts too . . . but actually he's a bigger believer today in it than I am."

Although members of the Ocooch Grazers Network became "believers," most of their immediate neighbors did not. One member explained:

There are groups in the state and in the communities that look down upon rotational grazers. I don't know if I should say this. Don't quote me. The Professional Dairymen's Association won't associate with anybody that's a rotational grazer because they are threatened by it. A lot of them have this tremendous capital investment in confinement systems, and they follow basically what the university told them to do, and they spent all this money, and they've done all this work. And a system that doesn't require that capital investment and the amount of work threatens them. And there's a lot more of that than you think.

To illustrate the extent to which rotational grazers were ostracized in the rural community, he went on to describe how difficult it was to find a "traditional confinement farmer" to serve on a certain local committee. Twenty conventional farmers refused before one accepted because they knew the committee included rotational grazers. Thus if network members did not cite technical drawbacks to adopting rotational grazing, they did cite social drawbacks.

Although many Ocooch members reported that the social climate in this rural area made the transition to rotational grazing difficult, they emphasized that the network played a meaningful role in buffering that problem. "When we first started the network, I thought it was just strictly to teach people grazing," explained Jim Brown. "But now, you know how alcoholics and gamblers have like a support group, I think I look at this as a support group. . . . People have ideas, and they can share them with people and tell if they're any good. You don't have anywhere else to take your ideas. I mean it don't wash." Likewise, John Hills elaborated on the idea of the network as a "support group" that helped him do something that otherwise would have been more difficult to do. Changing the way he had farmed all his life was challenging for John, but the network bolstered his resolve:

It's hard without reinforcement. I marvel at somebody that can just do it. But when your neighbors aren't doing it, it's hard. I mean you look around and nobody else is doing it. You are the Lone Ranger out there. And that's what the grazing network has helped me with. It's given me a little support, moral support. . . . Even if you know it's right, it's hard if no one else thinks so. . . . Is this like you're a bunch of alcoholics or something? You need somebody to help you not to take that next drink, you know, like AA [Alcoholics Anonymous] or something. It's like a support group.

Similarly, Vicki Braun appreciated the "camaraderie" that developed between the members: "I think that helps us personally just to get that support from other farmers . . . that have similar values in farming and that type of thing." As Charlotte Cannell put it: "They're learning something, they're sharing, [and] they're contributing also whatever story about whatever is growing or tried. They can look back now and laugh at something that happened. So the support is there through the information and the sharing."

The "support" and "camaraderie" that members reported finding in the network developed because of their shared interest in the technique of rotational grazing, which distinguished them from their conventional neighbors. Glenn Scoville thought grazers were considered to be "oddballs" by conventional farmers and by "people at the university," but at network events "you get to chum around with the people that's doing the same thing that you're doing, and they don't all think you're oddballs because if they did, they're an oddball too." Similarly, John Hills appreciated "just the getting to be with other grazers, and getting to visit with people that got things in common with you. . . . I think I have more friends with them than some of my close neighbors because they've got more in common with you than your neighbors do now." It was this shared understanding that led Tom McMahon to characterize those who practice rotational grazing as a "subculture" within the rural community.

While members of the Ocooch Grazers shared a common interest in rotational grazing, they did not necessarily share a commitment to the wider sustainable agriculture movement. During the course of this study, a shift seemed to be occurring with respect to the types of people who were adopting rotational grazing. Initially, the interest in rotational grazing and the organizational model of farmer networks emerged from members of the Southwest Wisconsin Farmers Research Network organized by the Wisconsin Rural Development Center. These farmers and rural advocates clearly and proudly identified with the sustainable agriculture movement. Some of the SWFRN members eventually became involved in the Ocooch Grazers Network; the geographic territories of the two networks overlapped somewhat.

Many of the farmers who first experimented with rotational grazing seemed to be part of what Mooney and Majka (1995) referred to as a cohort of young people who were college educated in the late 1960s and early 1970s and who returned to rural areas or wanted to escape urban life. Southwestern Wisconsin had been a popular place to locate because the hilly, marginal land meant farm prices were low. Many of this cohort tended to be activist oriented. For example, when the farm crisis escalated in the 1980s, they responded by forming rural advocacy organizations such as the WRDC, which not only promoted farmer networking but was also vocal in its criticism of the University of Wisconsin's research agenda, particularly with respect to the development of bovine growth hormone. At the same time, these farmers improved their own organic farming systems and created mechanisms for the cooperative marketing of their products. Given these and other initiatives, the adoption of rotational grazing was in many ways a logical extension of their fundamental commitment to sustainable agriculture. At an Ocooch pasture walk in 1995, one farmer shared the family "motto" undergirding their low-input, grass-based, organic dairy: "think liberal, spend conservative." For another family, rotational grazing simply fit within a larger vision of organic farming and "stewardship of the land."

Given this history, it may not be surprising that some members explained that the first group of people to practice rotational grazing in the area were considered by many to be "old hippies." For example, Jim Radke, the district conservationist, observed that the association between rotational grazing and organic farming continued to be made in the rural community and that association was generally negative:

Someone told me a comment that was made at these [grazing] meetings we had the other day. Two farmers were talking about organic farming, and someone mentioned that this farmer was an organic farmer, and he says, "You can't be, you don't have a ponytail." Organic farming has been stereotyped as a bunch of hippies, especially the early organic farmers, and that still hangs over. A lot of the sustainable agriculture grants and programs were done with what people perceived to be as a bunch of hippies. The intensive rotational grazing network is not a bunch of hippies, but they're a little bit more abstract than your normal, traditional dairy farmers, and I think you can see that by going to some of the grazing networks. More traditional people are getting into it now, but I think some of the early people were more abstract, more freethinkers than the others. They don't care about the traditional stereotype of shorter hair, the way you're supposed to dress, the way you're supposed to act, the way you're supposed to raise your kids. And I think you can really see that in a number of the organic farmers. And I hope I'm looking at this through open eyes, but I think that relationship is there. I know the perception is there among conventional farmers. I know it's one of the reasons we're not getting more regular farmers to jump into intensive rotational grazing because I think there's been some stereotyping of networks of about any kind.

Despite this stereotyping, Jim and others agreed that a noticeable shift was occurring within the grazing community.

Although Mike Cannell described this shift by saying that "grazing literally surmounted the sustainable agriculture movement" in Wisconsin, the word *surmounted* may be too strong. Those who identified with the sustainable agriculture movement continued to participate while the grazing community expanded to include those who did not necessarily so identify. As Tom McMahon described it:

The people that got involved in it [rotational grazing] when we did six, seven, eight years ago tended to be old hippies. They tended to be liberal, Democratic thinking, voting kinds of people. They were truly leftovers from the sixties and a subculture, and they got into it for a lot of altruistic reasons. . . . Initially the same people that were involved in rotational grazing were in the anti-bGH movement and in the Save the Family Farm network and involved in the Farmers' Hotline and all of that kind of stuff. And those people that were back then, still are [involved in rotational grazing]. But there's an awful lot of new people that are getting into grazing that are not that kind of an activist. I mean they're into grazing on their own farm, doing their own thing, and not involved in trying to shape the rural community.

What set this new group of grazers apart from the earlier innovators? These more "traditional dairy farmers" – to use the term that many network members used to distinguish them from the "old hippies" – were principally attracted to the economic benefits of rotational grazing as a technical system of milking cows. This was, for instance, Michael Hall's perception: "Everyone that gets involved in grazing is just doing it to try and better their bottom line number one, and maybe change in lifestyle." For some of these farmers, economic considerations were paramount because rotational grazing was a way for them to get out from under heavy debt loads and actually save their operations after going through bankruptcy proceedings. Some in the community tended to stereotype people with these economic woes, considering their management ability "suspect," as one farmer explained to me. Other newer grazers, however, were variously described as "real sharp," "professionals," and "better managers" who "are really into the economics of rotational grazing and into the economics of their dairy operation and for them it's purely a technical system of milking cows."

Whether they had been through bankruptcy or whether they were considered better managers, the newer grazers emphasized the economic logic of grass-based dairying above all else. Apparently, in the rural community, the early adopters of rotational grazing might have been written off as "old hippies" with "abstract" ideas. But when the so-called better managers decided to adopt grazing, it raised the ire of those who continued to farm conventionally, as Tom McMahon explained:

The smarter people and the better managers on the farms are beginning to see the economic value of it, and there are lots of people doing it that make a lot of other people crazy because they are. There's a place not too far from here that's got a great big confinement system, I think he's got six Harvestores and a big free-stall operation with a parlor and everything. Last summer he had all his cows out grazing. It makes other people with big confinement systems crazy that this guy thinks that it's more economically viable to abandon his facility during the grazing months and get his cows out on grass because the university says that you have to utilize that facility twenty-four hours a day, seven days a week, three hundred and sixty-five days a year in order to capitalize it or in order to spread the capital out. If you can go and abandon the thing for six to eight months a year and still be more profitable than keeping the cows in there, it says something about that system and what the university's been telling these people.

"I do see the grazers now as a group changing. We're getting more of the traditional, stereotypical dairy farmers into it," reflected Jim Radke. "People are using it now because they can realize they can still make money and still live easier and have healthier animals. Some of them are concerned about the fact that – and you hear this at the Ocooch Grazers all the time – that they're doing a much better job with the soil, which they are. But I think that's a secondary consequence of the first action of making more money for the group that we're getting now.'' Of course, the economic viability of their operations was equally vital to the early adopters of rotational grazing; however, the more traditional dairy farmers adopting grazing could be distinguished from those for whom rotational grazing fit within a larger vision of creating a more sustainable agriculture.

In addition to the idea that, as Glenn Scoville put it, ''the dollar is what drives it,'' was a tendency for some grazers to seem to want to distance themselves and rotational grazing from an association with the sustainable agriculture movement because of its negative image in the rural community. For example, grazer Michael Hall explained that the term *sustainable agriculture* meant organic farming to him. Michael was not an organic farmer himself, and he felt that ''a lot of people that are grazing are still raising corn and things like that and using heavy equipment and chemicals and fertilizers. I think we're partially sustainable ag, but that's not our goal. I don't think we've got a goal to be completely that way, or it's not something we're conscious of being or trying to do.''

Paradoxically, it was the network coordinator, Mike Cannell, who was the most adamant about disassociating rotational grazing from sustainable agriculture. This was curious because Mike considered himself a strong advocate of sustainable agriculture and was clearly the most activist-oriented member of the network. For example, he was a vocal opponent of bovine growth hormone, he enjoyed challenging conventional farming research when he served on numerous University of Wisconsin committees intended to promote further study of sustainable agriculture, and he was passionate and eloquent in his support of sustainable agriculture, making his views public through speeches and opinion-editorials in the newspaper. But Mike was clear about his perception of the need to disassociate grass-based dairying from sustainable agriculture:

The sustainable agriculture people would like, like a mother hen, to tuck rotational grazing under their wing, and say, ''Rotational grazing, that's sustainable, that's our baby.'' And they're trying to do that. There's a relationship that is created over and over and over among many people between grazing and sustainability. The fact is grazing is about the most sustainable way there is. It is a perfect example of sustainable agriculture, but whether the grazing community wishes to be aligned with the sustainable agriculture community is a philosophical debate and a political debate which to me is a different thing. . . . I don't think that the people who belong to the Ocooch Grazers should be forced to realize that they are sustainable agriculture–type farmers. There are people out here who think sustainable agriculture is some wacko, hippie, back-to-the-land, organic farming, vegetarian, pro-choice, you know, and they are not going to have anything to do with that movement. It's a movement. Among a lot of farmers, conventional farmers, this sustainable ag thing, organic farming, and all of that is just very, very marginalized philosophically. And if grazing gets hooked up with sustainable ag in that fashion, it will become even more difficult for these people to accept it as an acceptable technology. . . .

You know as a university researcher that grazing is tremendously sustainable. I know it is. But I never say grazing is one of the most sustainable types of agriculture because of all of this connotation that it brings with it when it walks into the room. . . . It's a gentle formation of credibility, and as the credibility builds and the acceptance widens, it becomes easier and easier for more and more people to adopt the technology, and bGH is the same way of course. But if grazing maintains this umbilical cord back to this sustainable agriculture philosophy, I just think it will stay marginalized to too great an extent in too many people's minds, even the extension people. If we can discuss the merits of grazing all on its own, as a way of taking care of the land and feeding the cows, it will be accepted much, much faster.

Mike's belief that rotational grazing ought to be disassociated from the broader effort to create sustainable agriculture was no less strategic than his more public activism. He was an activist embedded within this social movement community that, like other geographically localized movements (Stoecker 1995), included nonactivists. Ultimately Mike's goal was greater acceptance and adoption of rotational grazing among dairy farmers because of its sustainable characteristics, but to achieve this goal, he felt that he had to circumvent what was a "marginalized philosophy" among many farmers.

    Although there were strategic advantages, the distancing by some grazers of rotational grazing from sustainable agriculture poses a problem for the wider movement. Analysts and advocates have regarded rotational grazing as an excellent example of sustainable farming (e.g., Caneff 1993; Murphy and Kunkel 1993). Yet this association was a negative one in this community (and probably elsewhere). A pragmatic view of this problem is that it does not matter whether people *intend* to farm in ways that are more sustainable, as long they *do*. Echoing the National Research Council (1989), which speculated that today's alternative practices may become tomorrow's conventions, Sally McMahon felt that as more conventional farmers adopted rotational grazing because of its better profitability, the technique might become accepted more widely. "You picture conventional farming as the silos and our stereotype of a farm. But isn't conventional farming in other parts of the world more of a grass-based farming? Maybe someday down the road, there will be different types of farming, and it will be more respected. . . . Now grazing is an alternative. It'll become a choice." During this study, rotational grazing was considered by dairy farmers to be an alternative rather than a choice in the sense that Sally McMahon meant it.

Differences among network members in the degree to which they consciously identified with the sustainable agriculture movement were lessened by their pragmatic need for information and by their need for support from others who shared a common interest in rotational grazing. The loose organizational structure and the very specific and limited purpose of the network seemed to make it easier for members to tolerate different views of sustainable agriculture. In other words, because the group did not try to influence government policy and engaged in few outreach activities, they had no major decisions to make, and thus potential conflict with respect to explicit allegiance to sustainable agriculture was minimized. Members seemed not to have to negotiate their differences in ways that might have threatened their stronger sense of the network as a supportive community for farmers engaged in technical and social change.

Members of the Ocooch Grazers Network steadily gained confidence in their ability to generate understandings that they did not previously have, in other words, to generate knowledge. Network members recognized that the local knowledge they produced and shared with each other was the foundation on which grass farming would develop in their geographic area, even as they drew on other information sources. Interactions among the network members constituted not only the exchange of practical knowledge about how to graze but also ideological exchanges about why to graze. The network also functioned as a source of social support for its members who were disconnected from their neighbors continuing to farm conventionally. At base, a common interest in how to apply and develop the technique of rotational grazing was what held this social movement community together. As will become clear in the next chapter, this emphasis on technique and on physical place contrasts with the ways that knowledge exchange in the Wisconsin Women's Sustainable Farming Network was intricately tied to members' experiences in a gendered society.

FIVE

Personal Knowledge and the Creation of Women-Only Space

On a cool fall weekend in 1994, twenty-one members of the Wisconsin Women's Sustainable Farming Network gathered at a church camp located in Wisconsin's north woods. Members stayed at the camp's lodge, a large log structure with high, vaulted ceilings and a fire crackling in the huge stone fireplace in the center of the room. Towering white pines surrounded the lake and were visible through the picture windows. In this comfortable setting, members spent Friday evening sharing with one another lengthy "introductions" consisting of stories about who they are and how and why they farm. On Saturday morning, members led or participated in seven different workshops that covered a range of subjects related to farming and marketing. In addition to the usual introductions and workshops, the afternoon agenda included a large-group discussion led by a professional facilitator hired by the steering committee. With a large sheet of newsprint on an easel, the facilitator guided the group through a series of questions intended to draw out their personal goals and the obstacles they encountered as they tried to achieve those goals. These ideas were listed on the newsprint and then were used in generating a mission statement to express how the organization might help members reach their personal goals.

The formality of the group process used in the mission statement discussion was uncharacteristic of the network, which had been meeting periodically for over a year at that point. As a result, the discussion became somewhat strained; one member felt "frustrated by covering ground the group has already discussed," and another simply did not have "a lot of patience with that stuff," referring to the professional, "corporate" style of facilitation and generation of the mission statement. But the difficulties that emerged had to do only with the process of the discussion; no one disagreed on the basic content of the mission statement, which echoed many previous exchanges that had taken place among members:

Our mission is to inspire women farmers with a strong support network that promotes successful sustainable farming. We will share personal experiences, technical information, and marketing strategies.

In defining their organizational purpose, members of the Women's Network restated prominent themes that had emerged repeatedly in earlier discussions and that were subsequently confirmed in interviews. These themes and the primary functions of the Women's Network will be described and analyzed in this chapter.

The first section of the chapter presents an organizational profile of the network, exploring how the group constituted an informal social movement community. In the second section, I discuss the sources of knowledge that members drew on in their network exchanges, with particular emphasis on their personal knowledge derived from their social location and experiences in a gendered society, as well as on their local knowledge derived from the practice of alternative farming techniques and enterprises. The analysis put forth in the third section suggests that what members seemed to value most about the network exchanges was that women were communicating their knowledge to each other. That knowledge-exchange process consisted primarily of relating personal experiences of gender-related issues and experiential knowledge of technical information and marketing strategies. The sharing of personal experiences, a source of both inspiration and information for those who participated, helped members overcome personal and material challenges. In the fourth section, I review the ideological frameworks that undergirded the knowledge-exchange process in the network by exploring members' ideas, values, and beliefs about both feminism and agricultural sustainability. In the final section, I analyze how the group functioned as a "strong support network," an idea articulated in the mission statement and perhaps the primary function of the group from the perspective of its members.

Organizational Profile

Like other sustainable farming networks in Wisconsin, the organizational structure of the Women's Network can be described as that of a social movement community, that is, a loose association with informal patterns that characterize how the group functions (Buechler 1990; Stoecker 1995). The network did not always run as smoothly as some members would have liked, in part because it faced such challenges as the members' geographic dispersion, diversity of farming practices, and varying levels of agricultural experience. Despite the network's informality, members deliberately considered the potential implications of organizational decisions and occasionally experimented with the network's structure and activities. This organizational characteristic of careful consideration distinguished the Women's Network from other farmer networks with which I am familiar.

Membership

During this study, membership in the Women's Network was open and dynamic. The network's organizers intended the group to be principally for women who were currently or hoped to be farmers engaged in sustainable agriculture. In 1994, roughly two-thirds of the women on the membership list were actively engaged in producing and marketing agricultural products (e.g., food, fiber, forage, herbs, flowers). As will become clear in the course of this chapter, their levels of production experience varied considerably; many of the women were relatively new to farming, but a few had lived and worked on farms their whole lives. The group also welcomed women who were not actively involved in agricultural production but who were interested in sustainable agriculture for other reasons. As a result, in 1994 the membership list also included a considerable number of women who identified as either a home gardener, a researcher, a landscaper, a rural advocate, a government agent, or an agricultural journalist, among other occupations and pursuits.

Most of the women who organized the network lived in northwestern Wisconsin in the several counties surrounding the city of Eau Claire; however, many women from other parts of the state were soon interested in the burgeoning group. The 1994 membership list indicates the wide geographic distribution that developed within a year of the group's formation and continued during the period I studied it. The sixty-four members on the 1994 list lived in twenty-five (of the seventy-one) counties in Wisconsin or came from the neighboring states of Iowa and Minnesota. A regional breakdown of the membership list shows that twenty-six of the sixty-four individuals listed (41 percent) were from the northwestern area of Wisconsin. Another nineteen members (30 percent) lived in the southwestern part of the state. The remaining 29 percent of members included seven women from other areas of Wisconsin, four from Iowa, and eight from Minnesota.

The membership's wide geographic distribution led to one of the first organizational issues that the network confronted. Specifically, organizers grappled with the question of whether they wanted to have a local or a statewide organization. On the one hand, they valued the idea that organizing a locally oriented group could make possible more frequent contact with one another than if the group were a statewide organization. On the other hand, it was unclear to them whether there was a sufficient number of women in the immediate area who identified themselves as sustainable farmers to constitute a viable group. Most important, however, the organizers of the group were reluctant to exclude those who lived elsewhere and who actually constituted the majority of women who wanted to be involved in the developing organization. The immediate solution to this issue was to hold events near Eau Claire and to invite any woman who was interested in attending, regardless of her geographic location. Network members also set a longer-term goal, however, of helping women in other areas of the state to establish local chapters of a wider statewide organization.

During most of the period I participated in the network, from 1993 through early 1995, events were held in and primarily drew participants from northwestern Wisconsin, although several women from the southern part of the state did take part. From those who attended, I identified a core membership of sixteen women who regularly participated in network activities, and the interactions among this core group are of particular interest here. In March 1995, toward the end of my study, members decided at a business meeting to divide the network into two subgroups with the original cluster of women becoming a northern chapter and the newer cluster forming a southern chapter of the statewide network.

Steady growth in membership resulted from only minimal efforts to recruit new members. (From a March 1994 total membership of 64 individuals, the mailing list had grown by September 1996 to 286 women from around Wisconsin, as well as Minnesota and Iowa.) Major avenues by which potential members learned about the group included articles in the agricultural press, publicity for events, informational meetings held at the annual Upper Midwest Organic Farming Conference, and word-of-mouth. Joining the group involved simply getting one's name added to the mailing list, which was maintained by one member. An annual donation of five dollars to help defray costs of mailings was requested from each member, but there seemed to be little enforcement of this fee.

Leadership

The structure of the Women's Network included an effort to share leadership responsibilities among a small group of women who did much of the work of maintaining the organization. The somewhat malleable division of labor and leadership roles that characterized this structure were by design. In the

fall of 1993, six women developed a funding proposal, which they submitted to the competitive grants program of the Sustainable Agriculture Program housed in the Wisconsin Department of Agriculture, Trade, and Consumer Protection. Although the network had been under way for over six
months, some fundamental organizational decisions were made in the process of preparing that proposal and then later discussed with a larger group.
One key decision was not to request funding to pay a network coordinator,
as some other networks had done. The initial organizers of the Women's
Network were concerned that reliance on one or two paid coordinators
might limit the extent to which the general membership actively participated in and took responsibility for the ongoing maintenance of the organization. They reasoned that if funding became unavailable, lack of member
"ownership" in the organization could threaten the long-term viability of
the group, a problem encountered by another network with which several
members were personally familiar. Rather than rely on a coordinator, they
decided to have a steering committee that would plan events.

Although the Women's Network did not appoint a coordinator per se, Diane Kaufmann provided strong leadership. She organized the first gathering
of women as a special session of a larger 1992 conference sponsored by the
Western Wisconsin Sustainable Farming Network, described in chapter 3.
The success of that event inspired Diane, with the help of Denise Molloy, to
organize and lead the first of several meetings at which the idea of forming a
Women's Network was born. Diane was, however, more than the founder of
the network; she was active in other sustainable agricultural initiatives,
which gave her statewide name recognition. The familiarity, of her name as
well as numerous articles in the agricultural press associating her with the
Women's Network, led others in the sustainable agriculture community in
Wisconsin to make references to "Diane Kaufmann's Network." And, indeed, during the course of this study, she maintained an active role in the
Women's Network steering committee. Her commitment to the group might
have been derived in part from a sense of accountability implied by her status as the official contact person for the network's funding proposal, which
was eventually approved by the Sustainable Agriculture Program. But a far
greater reason for her commitment to the group was her belief in the unique
value of a sustainable farming network for women.

The leadership provided by Diane Kaufmann was complemented by that
of several other network members who helped do much of the work of the
organization. Although the makeup of the volunteer steering committee
shifted somewhat during the course of this study, four other women contributed much to maintaining the network. Perhaps the most time-consuming

task that the steering committee members undertook was the job of planning for network-sponsored conferences. Typically, the planning process involved the steering committee formulating a general outline of an upcoming event and then one or two members taking responsibility for planning that event in detail, recruiting presenters, and attending to logistical arrangements. Several members who were not necessarily on the steering committee assumed other organizational tasks such as maintaining the mailing list, acting as treasurer, and editing and distributing the newsletter. When the network was divided into two chapters in 1995, steering committees were established for each chapter, and four people in each chapter committed to one-year terms.

As originally intended by the network's decision not to have a paid coordinator, the organizational leadership structure allowed the work of maintaining the group to be shared between several leaders, and this seemed to foster a commitment to the group's success which might not have developed otherwise. Still, the structure presented some difficulties. Like many volunteer organizations, the flexible leadership structure depended on those who were willing and able to do work at a particular time. For example, during the course of this study, there were a few instances when steering committee members expressed to me their frustration that not all of the tasks people made commitments to do were completed. One explanation offered for this failure was that network meetings were intermittent. While energy levels seemed to be high during a meeting, maintaining a sense of accountability to the group was apparently difficult with infrequent face-to-face interaction. Similarly, despite leaders' attempts to encourage other members to help with certain tasks, such as planning the next meeting, new leaders did not step forward. As a result, several steering committee members expressed disappointment, albeit privately, that the work always seemed to fall on the same people. Perhaps not surprisingly, these concerns seemed to be most acutely felt by the steering committee members. Indeed, I never heard other members express awareness of this problem.

Anyone who has been involved with organizations will recognize that a loose division of labor may not always be the most efficient way to meet ongoing organizational needs. Indeed, these problems are not unique to the Women's Network, but they were more apparent than in the grazing network described in the last chapter. One reason may have been the relative complexity of the Women's Network as an organization because of its members' diverse characteristics and the range and type of the activities the group pursued. As a result of this diversity and perhaps other factors, the Women's Network required planning, coordination, accountability, and time from its leaders. My observations convinced me, however, that in comparison to other networks with which I am familiar, a far greater number of people were involved in the maintenance of this organization (e.g., tasks were delegated, women in the southwestern part of the state eventually took initiative to form their own chapter). Thus to a certain extent the initial goal of actively involving members in the maintenance of the organization was met.

Activities

The Women's Network pursued a range of activities. Later in this chapter, I provide more detail on the process and content of these activities in the course of analyzing the primary functions of the network.

Periodic conferences constituted the most vital and consistent, albeit infrequent, network activity. The first three conferences were one-day events held in a church's meeting room. In March 1993 the first conference drew sixteen women, a second one held in November 1993 attracted twelve participants, and sixteen women attended the third conference in February 1994. In addition, a day-long meeting of women who eventually founded the southern chapter of the network was held in September 1994 at a farm in southwest Wisconsin and attracted nineteen women. Subsequently, steering committee members decided to hold conferences over a weekend because they felt that the day-long events did not allow enough time to achieve all that they wanted to. A November 1994 conference at a church camp attracted twenty-one women, and a March 1995 conference at a lodge further south in Wisconsin drew forty-one participants. Child care was available at all of these events.

The conferences were multifaceted events, as the planned agendas allowed for a mix of both spontaneous and structured interaction. All of the conferences began with an opportunity for each participant to introduce herself to the group. Diane Kaufmann usually invited those gathered to "share who we are and what we are doing" and to "say something about what we want from the group." The members sat in a circle, and each woman was given an opportunity to speak. Often others questioned a woman further about what she said, and that frequently inspired other comments. These introductions consumed anywhere from one to two and a half hours, and no one seemed to mind devoting that much time to them. On the contrary, these introductions were, as Ann Hansen put it, "consistently . . . the most popular part of our conferences." Conference agendas also included sessions devoted to specific agricultural and organizational topics. At the one-day events, speakers made presentations to and engaged in discussion with the whole group. And the weekend-long events included both keynote speeches

As their mission states, members of the Women's Network come together "to inspire women farmers with a strong support network that promotes successful sustainable farming." Photograph courtesy of Diane Kaufmann.

and concurrent workshops from which participants could choose. Presentations were made and workshops were led by members and invited guests (both male and female) who had knowledge in an area of potential interest to members. Network business meetings were also part of the conferences, typically at the end of the agenda. Steering committee members presented ideas, and members discussed and made decisions on items such as the mailing list, funding, future activities, and the makeup of the steering committee.

In addition to the conferences, the network produced a quarterly newsletter. Network members decided to create a newsletter because of the geographic distance between members and the infrequency of conferences and other gatherings. The original idea was to have a member-written newsletter in which the women conveyed what they were doing, thinking, or learning about during the time between their conferences. Although the content and format of the newsletter changed somewhat over time, several women contributed to the informal publication, which was edited initially by Ann Hansen and later by Diane Kaufmann. Other mailings to members included detailed announcements of upcoming events. For members who were unable to come to events, the newsletter and mailings offered a way to keep in touch with others.

Another set of Women's Network activities took place on members'

A field day cosponsored by the Women's Network.

farms as work parties and field days. The suggestion to have work parties arose during an early brainstorming session on what activities the network might undertake. The idea was to allow any member to "issue the call," as Diane Kaufmann put it, and to invite others to help with a specific project and perhaps learn a new skill in the process. Although a few work parties were held during the period of this study, members seemed to view the idea as good in theory but generally unsuccessful in practice. When discussing why these events were few and turnout was low, several women offered explanations: not knowing which projects would be appropriate; feeling nervous about having people work on something that had to be done "just right"; and being unwilling to travel far to do something simple, like weeding, where the likelihood of learning a new skill was low. In addition to the work parties, three field days were held in 1994. All of the field days were open to the public, drawing both men and women, and two of them were cosponsored with the Western Wisconsin Sustainable Farming Network. At these field days, the host farmers explained their operation, gave a tour of their farm, demonstrated relevant techniques, and answered questions.

Although the Women's Network engaged in few outreach activities, one important avenue through which it gained new members was the Upper Midwest Organic Farming Conference (UMOFC), an annual event held in March that has attracted over six hundred participants in recent years. Although several members routinely participated in the UMOFC, Faye Jones was especially instrumental in maintaining a connection between the Women's Network and the conference. As a lead organizer of UMOFC, Faye scheduled an opportunity for the Women's Network to hold informational meetings as part of the larger conference. The purpose of these meetings was to introduce women to the Women's Network and to each other; as a result, the meetings resembled the introductions that began the Women's Network conferences. These meetings drew more than twenty-five women each of the two years I attended, several of whom eventually became active members of the network.

Linkages with Other Groups and Government Agencies

One early organizational connection was that between the Women's Network and the Western Wisconsin Sustainable Farming Network. According to Diane Kaufmann and Denise Molloy, key leaders in the SFN strongly supported the idea of forming the Women's Network and encouraged the women to apply for funding from the state. As the Women's Network began to organize, one woman suggested that perhaps the women's group might simply be part of the SFN. But Diane felt, and the other women who were the initial core membership of the Women's Network quickly agreed, that women ought to have their own group, rather than be, as they saw it, "relegated to a subgroup" or "a women's auxiliary" (a fairly common characteristic of other farm organizations, especially those that are commodity-based). Diane and Denise continued to be active in the SFN, but the relationship between the two groups was fairly minimal (e.g., cosponsoring a couple of field days) and intentionally loose.

Soon after deciding to form their own group, the organizers of the Women's Network applied for funding to support their efforts. Receiving funding from the Sustainable Agriculture Program constituted the most enduring and significant organizational linkage that the Women's Network had during this study. The proposal, submitted in the fall of 1993 and titled "Skill Development for Women in Sustainable Agriculture," requested $9,000 for three years of network activities. After the review process, the Sustainable Agriculture Program chose the Women's Network as one of nineteen projects to be funded (out of a total of fifty-five proposals) beginning in 1994 and as only one of two networks that received funding in that year. According to program staff, some Sustainable Agriculture Advisory

Council members, who voted on which projects would receive funding, had been opposed to granting money to the Women's Network. Those opposed reportedly raised questions about why women needed a separate organization, and some even wondered whether it was legal to grant money to a gender-exclusive group. One of the factors that apparently increased support for the proposal from the Women's Network was that none of the money requested was going to be spent on salaries for network coordinators and that instead the money would cover direct costs of distributing information to network members. Ultimately, the Women's Network was awarded funding by a five-to-two vote.

Financial support from the Sustainable Agriculture Program was an important resource for the Women's Network during the time of this study. Because the network had already been under way before receiving funding, members claimed that they would have continued with their efforts even if the grant request had been denied. Nonetheless, funding clearly made certain network activities possible that might not have been easily achieved otherwise. For example, organization building can be a slow process, as the low participation rates at early events seemed to indicate. Funding enabled the women to hold weekend-long conferences which enhanced the network's ability to attract more participants, to develop a solid membership base, and potentially to strengthen the longer-term viability of the group. In addition, the Women's Network received a publicity boost thanks to the large mailing list maintained by the Sustainable Agriculture Program and its practice of distributing to everyone on that list a calendar of events sponsored by its grantees.

Developing Personal Knowledge

One of the purposes of this study was to consider the ways that different lived experiences might produce multiple and partial perspectives from which local knowledge for sustainable agriculture is generated. In the rotational grazing network described in the previous chapter, members clearly exchanged ideas that had been derived largely from their local knowledge about the particular sustainable farming technique they all used. By contrast, members of the Women's Network pursued a range of farm enterprises and engaged in a variety of farming practices. Rather than a shared interest in a common technique, members of the Women's Network were united primarily by a common social location, specifically their gender identity. That positioning was a principal source of personal knowledge upon which members drew in their network exchanges. From their own personal experiences in a gendered society, women in the network developed certain ways of seeing, knowing, and understanding farming experiences which seemed to differ from those of their male counterparts. Carolyn Sachs (1996:17) made a similar observation about rural women more generally: "Rural women's knowledge is situated in their particular localities and daily activities. . . . These experiences provide particular angles of vision or partial perspectives that offer the possibility of seeing differently than from dominant perspectives."

Network members' experientially based observations derived from their gender identity were a prominent and recurring theme during network activities and in the way members characterized the network in interviews with me; therefore, the discussion that follows begins by exploring network members' personal knowledge about gender relations in agriculture. Another distinct theme, however, stemmed from the fact that network members were not only women but also farmers. That is, although these women did not engage in a common farming practice as the members of the grazers' network did, it was a network for women interested in sustainable agriculture. Not surprisingly, then, local knowledge about sustainable farming practices and management was also an important, though secondary, basis for network functioning. This theme will also be examined in some detail below.

Social Location and Personal Knowledge

In observations of network events and in individual interviews, members recounted numerous instances of their personal experiences with gender stereotyping and gender relations in agriculture. Specifically, many of them recognized the general societal assumption that farmers are and should be men, not women. One result of this gender stereotyping was that many of them had gotten the message that it is not their place to be farmers. Network member Wendy Everham, who had recently moved from a large urban area to rural Wisconsin with the hopes of transitioning from a longtime home gardener to a market gardener, recalled her first exposure to the Women's Network: "I really think I was shocked to find that women farmed."

That shock reflects the messages society sends to women about the possibility of their farming. As recounted in chapter 3, Diane Kaufmann felt as a young girl that farming simply was not an option for her because of her gender, and, despite her strong interest in attending an agricultural college, she kept the dream to herself, afraid that she would be laughed at by others. She recognized that "society has not really allowed that desire to farm to surface, or encouraged it, or in some cases put up so many roadblocks that it wasn't possible." Diane told a similar story about her grandmother:

She wanted to farm that place, the homestead, and it went to the oldest son. And it kind of created some hard feelings as I understand it. It's just the way it's been forever, I think, is the oldest son or a son inherited the farm, it was never the daughter. So it was like even society doesn't encourage women farmers. . . . There's definitely some kind of stigma against women wanting to be involved in being a farmer. You can be a farmer's wife, and that's a role that's well understood. But to be the farmer is, you know, "Why would you want to do that?"

Another member, Jo Bauer, told of learning at a young age that women are discouraged from becoming farmers. She recalled strong childhood memories of not being allowed to participate in activities related to farming:

My grandparents on my dad's side did have the farm. . . . When I went up there, us girls we couldn't ride in pickup trucks. . . . This was back in the early sixties. I was little. I remember one time they were going to go pick up peas from the pea factory to bring home, and they could be canned then. I couldn't go along because I was a girl, and I cried. And you were discouraged from going to the barn. You were discouraged from going down to the creek and playing around. Girls didn't do that. Girls stayed in the house and cooked, very helpful skills, but just not what was interesting.

A perhaps dramatic example of assumed gender roles in agriculture came from member Nan Smith, who reported being told by a high school guidance counselor that she "must be gay" because she wanted to do something only a man does, that is, to farm.

That members received negative messages about wanting to farm is somewhat paradoxical because in every period of agricultural history women have actively participated in agricultural production around the world (Neth 1995; Whatmore 1991). Social scientists speculate that when human societies first made the transition from gathering and hunting to agricultural and nomadic ways of life, it was women, not men, who discovered how to domesticate and cultivate plants by applying the knowledge they accumulated through their food-gathering activities (Rosenfeld 1985; Shiva 1989). In her extensive collection of historical documents, *With These Hands,* Joan Jensen (1981) chronicled a rich and diverse history of women working on the North American landscape, a history of women's experiences that has often been ignored or underappreciated until recent decades. This history includes the indigenous women who cultivated corn and beans long before contact with Europeans; Sojourner Truth, who spoke in 1851 for other women who had been enslaved when she countered the argument that women were frail: "I have plowed, and planted, and gathered into barns, and no man could head me – and ar'n't I a woman?" (quoted in Jensen 1981:57); a half million farm women who engaged in agrarian social protest and mass organizing during the tumultuous period from 1870 to 1940; Euro-American women who have provided labor essential for the survival of family farms; and campesinas who have tended cotton in the Texas sun under harsh working conditions and who have sought bargaining power through the United Farm Workers. Women's participation in agricultural production shows tremendous variation across space and time, but women are most certainly not newcomers to farming.

Although women are not new to farm production, their roles in agriculture have changed over time, and members of the Women's Network had chosen to farm in an era that was profoundly influenced by the process of agricultural industrialization during the nineteenth century. Recent scholars have documented what network members themselves know from experience. That is, farming has come to be seen as a male occupation, while women's work in farm production has been obscured completely or considered secondary and supportive (Haney and Knowles 1988; Sachs 1983). This gender division of labor is evident on family-owned and operated farms, the most prevalent form of production in Wisconsin.

Historical works by Mary Neth (1995) and Katherine Jellison (1993) have documented that the research and extension activities of agricultural colleges and universities helped to shape these changes in gender relations on family farms. Beginning in the first decades of the twentieth century, agricultural professionals from land-grant universities disseminated the fruits of agricultural research to rural men, while the results of research into domestic science or home economics were disseminated by female professionals to rural (and urban) women. Although early home economists often had what they were convinced were women's best interests in mind, they articulated a domestic ideology that established a set of behavioral norms defining the ideal role of women as homemakers. Reinforced and promoted through educational texts and the popular media, men's work became judged by new, professional standards of production efficiency. Women's household labor was slowly separated from the productive work of the farm enterprise and judged by class standards of consumption, that is, the degree to which a household bought consumer products. Although many farm women resisted efforts to release them from their participation in farm production, ultimately gender relations and the work cultures of farm men and women were redefined, making a clearer demarcation between the farmer and the farmer's wife.

One apparent result of the gender stereotyping in agriculture is that women's participation in farming activities is often rendered invisible. For instance, one member described her observation that her knowledge and participation in her family's farm were devalued in various settings because of her gender:

Women get a lot of different things just from being a woman which is very typical. . . . Just the automatic assumption that you don't know what you're talking about because your husband sent you to the feed store to pick up the grain. Or your husband sent you to the implement dealership. Or you go to the courthouse to sign papers on getting water diversions, and they want your husband's name, and they want you to sign your husband's name. And I said no, I'm the one doing all this; I'm the one. My name is on the farm too, I'll do it. And it's just not a given.

Likewise, Wendy Everham explained that these experiences can happen in "subtle ways" that can still be "pretty intimidating." She cited the instance when she went to buy a tiller for her market garden, and rather than talk to her, the salesman kept addressing her husband, who had gone along: "Finally, my husband said, 'Well, I'm probably not going to be using that. You should probably be talking to her.' And he still couldn't. If the two of us were standing side by side, he still wanted to talk to [my husband] about all the attributes about this tiller." Similarly, Denise Molloy explained that in many settings, such as farming conferences, she has not felt that she "belonged there" and "even me going down to my local feed co-op, I always feel like they look at me like I'm from Mars. . . . I don't know if it's just our specific area, but most of the time I don't feel really treated with respect, that's just been my experience." And as Faye Jones found: "Nothing is built, gloves, equipment, everything is not built for a five-foot-two, hundred-and-twenty-pound person. It's built for a five-foot-ten or six-foot [person]. Tractor seats, everything. All those things are a pain in the butt. There is only one place I can go buy gloves that fit me properly, and I pay more for them even though they're smaller. I pay 40 percent more than the man's jersey gloves, but they fit me right." Similarly, Winnifred Moths (known to her friends as "Freddie") wondered: "Boots and gloves aren't made for women. Don't they think women work?"

Other members experienced having their farmwork obscured or taken for granted because of their gender. For example, Jo Bauer reported observing in her everyday life how women's contributions were devalued in a male-dominated culture. Jo and her husband, Doug, were dairy farmers and had an agreement to divide the work between them so that he took care of the cows, and she raised the field crops and maintained the basic farm machinery. In addition, she raised a market garden. But they lived on Doug's family's farm, where he grew up, and as she put it, "This is his farm, and it always will be, so even though we have this agreement it's not always held to." She explained more fully:

It's really hard being female in a man's occupation, or what's been traditionally a man's occupation. . . . I get tired of the jokes sometimes, so I tell them . . . I mean there's some people that, oh, they'll sit there, you know, two guys yakking away, "Oh my wife, she's always got to go here and there, and all they do is gossip," and I'll sit there going "Oh really, and what are you two guys doing?" It's just the women aren't valued for what they do is what it comes down to. A lot of what they do is taken for granted, and I'll get mad at Doug sometimes because I'll say, "Wait a minute. Stop. I did this, this, this, and this for you. Now what exactly have you done for me?" And he'll just say, "Well," and he'll admit he can take a person for granted pretty fast. And it's easy to do when you're not writing out a paycheck to somebody.

Similarly, Freddie Moths, interviewed when she was in her mid-sixties, had seen the way women's contributions were minimized as she grew up on a farm. Then, as an adult, she directly experienced her own labor being devalued when she and her husband, Reuben, operated a dairy farm in southern Wisconsin before their barn burned down thirty-five years ago. Thinking back to those days, Freddie recalled:

Women have always helped on the farm; they've always been out there in the barn milking, in the fields, they've always done that. . . . The women were out there helping, yes, some more than others. Of course, some I knew, well they never went in the barn, and then the others were out there helping with everything. But it was the man running the farm, making the decisions, you know, and the pocketbook. . . . [Women] were labeled just as the farm wife. They were the wife. . . . I always said I was the unpaid hired man. And I think that's the way it was for an awful lot of women. They worked very, very hard, it was their family's livelihood, and if they was a success, it was the man who took the honors. It was the man who made the decisions, it was the man who wrote the checks. When the farm was sold, and it was written in the paper, how was it written? Or when the man died, "The man was a very well-known farmer in the county" and so on. But the woman died, what did they say about her? They didn't say she was a well-known farmer that did this or that in farming. They never mentioned her accomplishments in farming.

When the Moths's barn burned, they sold their dairy farm and moved to a small farm in northern Wisconsin. Reuben took a job in a factory, and after her children were grown, Freddie said that she found herself "twiddling my thumbs." She decided to raise sheep and lambs for market, but eventually the business evolved into a focus on fibers or producing high-quality dyed yarns and wool which she sold primarily to spinners. But the difficulties she faced as she began to develop her own business revealed that power differentials based on gender can have debilitating effects:

At first, I think I was very insecure about what I was doing and who I was and very shy. When I started with this business, I had a terrible time to get myself to even call someone on the telephone, or to talk to someone on the telephone was very, very hard for me to do. And now I do it and think nothing of it. . . . My education ended when I graduated from eighth grade. I was thirteen years old. And from that time on I was just home on the farm, raising the children, doing the farmwork. First at home with my folks, and taking care of my little brothers and sisters, and my mother worked out, and I had to keep house. Then it was marriage and into that home, and not ever working out, never having my own money. You really just never grow. Anybody who's never been in that position cannot understand that, can't even fathom this, but here you are thirty-five years old still getting an allowance, as if you were a little kid. And you don't grow. You just don't grow mentally. You're an adult, but you never make adult decisions.

Perhaps as debilitating as not having any control of one's own finances or farm decisions was the expectation encountered by some women that they must do all the housework and child care, as well as contribute to the family income. Faye Jones, who at the time of the interview was a single mother in her mid-thirties, reflected on her previous relationships with men and how they affected her farming:

I have had to slowly come to the realization for me . . . [that] the husbands I've had or the men in my life have greatly taken away from my work in agriculture. I mean many people assume that I got this farm when I was married or that my husband had the tractor, and it's quite to the contrary. And I think it had to do with what some of these guys expected from me as a woman. They not only wanted the wife, mother, housekeeper person, but they wanted me to totally support myself and contribute to the household income, and you can't do it all. At all. You end up tired and unhappy. . . . They would mouth-talk really, oh they thought it was great, but when it came down to watching the baby for six hours, four days in a row, so I could really get something done, that was a different story. . . . Whenever it came down to meals or child care, ultimately it was my responsibility. So I didn't have a choice because these children had to eat every day, and they had to be taken care of.

The struggles network members personally encountered as a result of gender stereotypes and relations in agriculture led them to understand that they operated in a context that discouraged their farming. At the same time, however, they did not necessarily see these limits as all-encompassing or all-powerful. Rather, an important element of some network members' personal knowledge seemed to be their own understandings of themselves as individuals who could make choices and take actions on their own behalf despite social constraints. This sense of personal agency became clear in the network exchanges discussed below, and it was also referred to by certain members in their interviews. For example, market gardener Jean Schanen said: "There are obstacles of acceptance and getting the male world to pay attention to you . . . that I think other women tend to experience more than I do because of my weird background. . . . I was kind of raised to be a man. My father really, really wanted a son and didn't get one . . . and so I was raised to be a woman in terms of housecleaning but a man in terms of expectations in the world. I just kind of grew up thinking I could do whatever I wanted to do." Although Jean and several other members reported having been raised with confidence in their own abilities, Freddie Moths described the slow process by which she developed her own sense of agency when she began to experience success in her sheep enterprise:

I was just always very painfully shy about talking to anybody, expressing myself, expressing an opinion. I didn't feel I had a worthwhile opinion to say. That's just the way it was. And that has only changed now in the last fifteen years, simply because I just slowly got in with the sheep. I learned something, and I realized I knew something that other people could come to me and ask advice about. I started to get a self-confidence about myself. Learning my art, with the dyeing and the blending and stuff, I started to realize I had some self-worth too. My opinions counted. My choices counted. You slowly start to realize this. With just a little bit of success, it kind of goes to your head. . . . You're not afraid to talk on the telephone, you're not afraid to meet somebody new. . . .

Now I make decisions on things, I decide how and what to do with the sheep, when they should be wormed and all those things. And I have my own checkbook, and I decide whether I can afford this. I mean I don't have to ask someone else. . . . It is a big, big difference in a person's life. . . . It has added a lot to my life because take that away, and I'd be an entirely different person than what I am now today.

Lois Nerby, in her mid-eighties, also had a strong sense of her own agency and commented about her attitude toward men: "I never felt, myself, that they were any better than I was. If you can do it, I can do it." Lois said, "I've been in the middle of agriculture all my life," and she recalled being allowed to learn agricultural skills along with "domestic economy" through the 4-H and through a county school for farm teens. At a young age, however, her sense of agency was tempered somewhat by the experience that less was expected of her because she was a girl; for example, she felt that in 4-H she always had to "prove" that she was "better than the boys." Likewise, Carrie Mann reported, "I find myself sometimes feeling as though I need to prove myself." And Wendy Everham made a similar observation about women in agriculture: "In a world where it's a man's job . . . I still think women have to do things better than men, twice as good as the man, to get any kind of recognition."

No less than the grazers whose farms differed from one another and who adapted their technical knowledge to a particular place, members of the Women's Network also had lived different experiences despite the patriarchal gender relations that pervade society more generally. The network included women with husbands, women with lesbian partners, and single women. It included women with considerable wealth and others who must watch every penny. It included women from a variety of geographical regions in the state. It included women in their twenties and women all the way up to their eighties. It included women who were new to farming and women who had farmed for more than fifty years. Accordingly, their personal, experiential knowledge of gender roles and relations in agriculture varied considerably. Such diversity provided a rich source for knowledge exchange. But there was also a common thread running through their personal knowledge: a societal message that women are not farmers and are not capable of farming.

Local Knowledge and Sustainable Farming Practice

Network members were united not only by gender but by a common interest in sustainable agriculture, and one of the main activities of the network included the exchange of technical information about alternative farming practices. The sources of knowledge on which network members drew for these exchanges varied to some extent because members had developed or were in the process of developing farming and marketing enterprises that differed from one another considerably and these differences in enterprises meant that the various farmers necessarily employed different techniques. Moreover, even on particular farms, some practices used were those associated with sustainable agriculture (e.g., rotational grazing, organic vegetable production), while others were not (e.g., using synthetic fertilizers). Nonetheless, analysis of the data indicated that several generalizations about the sources of knowledge relied on to pursue those practices associated with sustainable agriculture can be made.

In describing how they learned about alternative farming techniques,

Women's Network members stressed the value of "observation," "experimentation," "just actually doing it," and "trial and error." In other words, they generated local, personal knowledge just as did members of the grazers' network. As Jean Schanen put it, "I can't say all, but a whole lot of our useful information comes from our observation and experience." Carrie Mann's experimentation on the basil farm was "evolving all the time . . . because I feel like I'm sort of really just beginning to feel confident in my farming abilities, so I am more able to understand what is needed and what kinds of experiments I can do to see what really works. So, like this year, we're going to do several rows of planting clover, and then we'll hoe out where we want to plant, and we'll plant right into the clover. Where we farm is really sandy, so we're always trying to do things to build the soil and just keep it from blowing away." Similarly, Jo Bauer experimented with cover cropping in her market garden and closely observed the soil: "I actually learned a lot more about the soil on this farm two years ago when I started planting the [market] garden because I was in there with my hands, and we covered quite a bit of ground, and you get to realize the differences within ten feet. You can see it on the tractor, but it is a much sharper observation when you're in there with your hands. . . . I can tell you where every wet spot is on this farm. I can tell you where every gopher hole is on this farm. But until I did the gardens, I couldn't tell you exactly what the soil felt like." And Faye Jones summed up her own knowledge development: "After a while you just, you start to develop an intuitive sense on how to do things."

Even when network members learned about ideas from other sources, they reported that it was important to them to figure out what worked best in their particular situation. Diane Kaufmann offered an example based on her experience of developing an operation that includes milking sheep. Although sheep dairying has a long tradition in other countries, it is a very unusual enterprise in the United States. To feed her sheep, Diane relied principally on intensive rotational grazing and tried to minimize the purchase of forage from off the farm. Recently she incorporated into her flock some East Friesian genetics, which is one of the best European breeds for sheep dairying. But the novelty of sheep dairying and the needs of this particular breed mean that Diane has much to learn through experience:

Trial and error, I guess, is certainly part of it. . . . Kind of a case in point now is feeding these sheep. I've been feeding my sheep one way, and now as I go to this dairy sheep symposium and hear how this breed of sheep, Friesians, that we're working with are going to require a much higher plane of nutrition probably than what we're used to feeding these sheep to get milk out of them. So now I'm into this

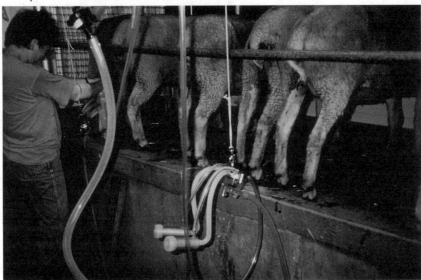

Members of the Women's Network are engaged in a range of alternative farming practices, from cut flowers to sheep dairying.

debate with myself. Can they do it on grass or do they have to have this higher plane of nutrition which will require buying more feed? And so it's like you get so many answers out there. You get more answers than you want to hear, and I guess what I've learned is you can hear a lot of variations on how to do something, but in the end you have to bring it back to your farm and say: these are my resources, this is where my comfort level is, this is what I'm going to do, and see if it works. And then hopefully evaluate your results. . . . One of the bigger lessons I've been learning is that every place is different and what's going to work for you here isn't going to work necessarily for someone else, and you've got to be the one who finally makes the decision about what's right for your place. And I always kind of probably wanted cookbook answers before too. And now I have to decide what fits your philosophy, your way of doing things, and live with ourselves.

Diane clearly recognized the importance of her own action and reflection process, resulting in her local knowledge about what works best in her particular situation.

In addition to learning from their own experience, network members reported learning technical information from other farmers' experiences. Carrie Mann summed up gathering information about organic farming: "Other farmers, generally speaking, have the best practical, you know, tried and true information." For several network members, learning from other farmers also included being hired or serving as an "apprentice" on an organic farm to gain technical skills. In addition to informal conversation with other farmers, many network members mentioned learning from other farmers at conferences, such as the Upper Midwest Organic Farming Conference, and at events, such as field days sponsored by sustainable farming networks where "going to people's farms and seeing how they do it" proved valuable, as Denise Molloy put it. Faye Jones recognized that farmers in alternative agriculture drew on each other as a resource in part because the information was not available through conventional means: "I think that farmers sharing information with farmers is, particularly about organic, sustainable farming practices, something that is happening because we can't go to our extension agent. . . . If you don't just know [something], talking to your neighbor or another woman is one of the best ways."

Although network members stressed that learning from other farmers with direct experience using alternative farming techniques was especially valuable, they did not reject other sources of knowledge. Rather, like the grazers who occasionally invited scientists to speak at their network events, the women were willing to learn what they could from both men and women outside of the network who could share information that members wanted. For instance, Diane Kaufmann described her decision to invite a man who was a leader in the grazing movement to speak at the first women's network gathering:

I felt he really had a story to tell women that could give women a bigger picture of what they could be doing with grazing on their farm. And I didn't care that he was a man. He had a story to tell that I wanted other women to hear because this kind of grazing was so low input and we can do it kind of thing. So it was kind of a desire on my part to spread the gospel, the word. But I wanted it in a situation, in a setting that women are going to be comfortable coming to, and in that process we'll discover whether it works or doesn't work, if it's valuable for us to come together. And I think what we found right away is that yes it does, that there is a different level. We are comfortable doing this.

After that first event, the Women's Network continued to bring in non-farmers to lead workshops or sessions at their conferences, including a government official who compiles soil surveys for the Natural Resources Conservation Service, a woman carpenter, and a man who sells and installs fences. Although most invited guests covered technical topics, at one network conference I attended, Kate Clancy, a nutritionist and leading scholar of sustainable agriculture, addressed gender issues in a talk titled "Women and the Food System." Thus, although network members affirmed the value of the personal knowledge they had to share with one another and recognized their need for women-only space, they were also willing to learn from others, albeit in an atmosphere that they found comfortable.

Despite a willingness to learn what they could from the dominant system of information distribution, none of the network members mentioned research and extension activities associated with the university as an important source of knowledge. Moreover, members responded negatively when I probed further and asked whether they had found the university to be a useful source of information. Some members indicated that the problem was that the university's research was conducted on a much larger scale than their own farms. For instance, although Denise Molloy attended "sheep days" at the extension research station in her area, she found that the information provided was not very relevant to her own situation:

They have sort of unlimited resources, financial resources. They always have these like gorgeous facilities, and everything is just set up perfect, and they have all the space they could ever want and all the equipment. . . . They don't struggle with the same having to make a profit that a regular producer does. . . . If you don't have somebody else paying for this wonderful setup and you've got to justify your expenses by how much money you're making off of your sheep, then you end up having to find other ways of doing it. I'm really small scale, and just from observing their facilities it is kind of like way beyond what I could ever afford.

Others found the extension services to have little or no information on alternative farming practices, such as Jo Bauer, who said of her county extension agent, "He's done a terrific amount of work for the dairy farmers in this county, and yet he's the one place you really don't go for organic information." Likewise, organic producer Faye Jones had not found the extension services in her area to be useful: "I think there's been a change, but I don't even make an attempt to contact any of them with my questions. Whenever I have, they've known so much less than I already knew. People call me now. . . . When people call the extension, you know what they do? They give them my name. It's really, really weak, and I think it's slowly changing. . . . But I've never been successful in utilizing that source because they don't know because they weren't taught." Organic basil producer Carrie Mann had limited experience with the university because "a funny thing about growing basil is that we're more experts on basil than a lot of the people that we are asking questions of because they simply don't have the experience with it as a product." Although she praised a university entomologist who helped her identify insects found in her basil field, her most in-depth experience with the university, a week-long class on integrated pest management, or IPM, was less satisfying:

> It was fundamentally really informational but really disappointing to me. Because at the time that I took the class, it felt to me as though the university viewed IPM as informing the farmer better about what is going on in their field before they spray in order to use the right type of spray and the right amount. I was looking for something a little different. I was sort of thinking of IPM as like you spray as the last resort, absolutely like you stop praying and start spraying. And it didn't feel like that at all to me. They mention all the chemical things by their brand name. I mean it's like everyone nodded in agreement like they knew what they were talking about, and furthermore they all sounded like missile names to me. You know they're called like Patriot and just some weird things.

> So that was my criticism of it at the time, though there was a lot of great information, it just totally fell back on what they call a conventional farming view where you identify the problem and then you get out there with the spray. They definitely emphasize scouting the field and knowing what's going on, but they never discussed any alternative types of solutions. I mean they never once in an entire week of being in that room did they ever say the words *cover crop*, like as a possibility to help remove some fungus from the soil or to help attract some beneficial insect.

Thus, whether because of the relatively small scale of their operations, because of the organic practices they used, or because of the uniqueness of their product, network members generally found university research and extension services to be of minimal utility. Instead, they tended to rely on their own local knowledge and that of other farmers in the sustainable agriculture movement.

Besides their own and other farmers' experiences, the only other major source of knowledge that network members mentioned was publications. For the most part, specific references to books and magazines reflected the practices that members used. For instance, those who practiced intensive rotational grazing mentioned the same books that the grazers described in the last chapter used, such as Bill Murphy's (1991) *Greener Pastures on Your Side of the Fence*. Several general resources that related to sustainable agriculture were also mentioned. Among those most frequently cited were the *New Farm Magazine*, formerly published by the Rodale Press, and *Acres USA: A Voice for Eco-Agriculture*, edited by Charles Walters Jr. Another source mentioned by four of the eleven members interviewed was materials provided by the Appropriate Technology Transfer for Rural Areas.

This federally funded program (as described in chapter 2) answers specific requests for information on low-input and other alternative farming practices. While publications were cited often by these women, some of them emphasized that these knowledge sources are limited. As Ann Hansen put it, she needed to develop "farm-specific information. . . . You can only learn so much from books and other people. In the end, you have to just go out and dink around and make mistakes." In sum, we see a central theme similar to that identified in the analysis of the grazers' network: a heavy reliance on local knowledge for farming techniques.

Exchanging Personal Knowledge

The mission of the Women's Network was, in part, to "share personal experiences, technical information, and marketing strategies." The term *personal experiences* was deliberately added to the mission statement because network members felt that the sharing of technical information and marketing strategies did not adequately capture all that the group was about. As Denise Molloy pointed out, "Even though the members are all interested in farming better, the network seems to focus more on being women." Indeed, of the interactions I recorded that constituted exchange of personal knowledge within the network during this study, roughly two-thirds dealt with gender-related issues. By knowledge of gender-related issues, I refer to my observation that members of the Women's Network exchanged what they called personal experiences of a social problem, or, more specifically, knowledge derived from their encountering gender inequity and overcoming those obstacles that they faced as women farmers. The remaining one-third of knowledge-exchange interactions that I observed dealt with subjects relating to the technical aspects of agricultural production and the marketing of produce. Although exchanges about production and marketing were not the primary function of the network, such exchanges were nonetheless an integral dimension of how the network functioned for knowledge exchange and an aspect that became increasingly important over time. In the discussion that follows I begin by exploring the gender-related issues before turning to the exchange of technical information and marketing strategies within the network. Last, I discuss certain characteristics of the knowledge-exchange process.

Personal Experiences of Gender-Related Issues

If identifying as a grass farmer constituted a technological reversal from the dominant methods of dairy production, a woman identifying herself as a sustainable farmer constituted a social reversal. No less than the grazer who needed knowledge to make that technological reversal possible, so too the members of the Women's Sustainable Farming Network needed knowledge to make that social reversal possible. Because knowledge is generated from a particular and partial perspective, the content of the knowledge exchanged among network members was shaped by their social location. This is not to say that all members held common views. Rather, as the sociologist Dorothy Smith (1987:78) argued, what women "have in common is the organization of social relations that has accomplished our exclusion." In contemporary agriculture in the United States, women have been excluded in ways that devalue their knowledge and labor and that fail to recognize their identities as farmers in their own right. Members of the Women's Network shared this "standpoint" (Smith 1987), which was evident in the distinctive characteristics of their knowledge exchange. The content of this knowledge generally related to certain institutional impediments and social constraints that affect women farmers and that are often deeply embedded in rural communities.

Tracy Ehlers (1987) documented the biases that often emerge when women deal with lending institutions, implement companies, feed stores, stock dealers, and the like. Network members stated this problem clearly in their grant proposal: "Obtaining necessary financing, supplies, and technical assistance is frequently difficult or impossible for women farmers simply because of prejudice against our gender." In addition, network members encountered certain challenges that were less the result of institutional barriers than of their own socialization and the unequal gender relations they experienced in their personal lives. Although not all members encountered the same difficulties, the examples below show that at network events members shared their personal experiences, and in this way they educated one another about these problems. Perhaps more important, they shared specific ideas on how those obstacles might be overcome.

Difficulties obtaining financial credit seem to be the most common institutional form of gender inequity confronted by women farmers generally. (Ehlers 1987). Beginning farmers often find financial assistance difficult to obtain, and when these farmers are women, bankers are even more skeptical. The extent of network members' personal knowledge of this subject varied. While some women struggled with their credit situation, others had enough money of their own and therefore did not experience direct difficulties in access to credit. Through network interactions members learned from the experiences of others. For instance, the credit issue arose at one of the first conferences organized by the Women's Network. At the time, Faye Jones was having difficulty refinancing her farm; and although she acknowledged that not all of her credit problems arose because of her gender, she observed: "I know if I had been a man it would be a different deal. . . . I do think women have strikes against us when we go into the bank." Faye asked others for any information they might have that would help her. In response, Freddie Moths shared her own experience of circumventing a loan officer who refused to consider her application by having her husband borrow the money in his name. That was not an option for Faye, a single mother, who felt that her positive credit history had "disappeared" when she divorced. Others suggested that there were ways to build credibility with lenders, such as preparing detailed business and tax plans. In response to the need for information in this area, the organizers of the next network meeting dedicated much of the agenda to sessions called "Business Planning" and "Tax Planning for Farmers."

In addition to such institutional challenges, members of the Women's Network needed certain skills that they felt they had not developed because of gender discrimination. For example, the network's grant application stated: "Women farmers often do not have the mechanical training and interests necessary for using and maintaining tractors, implements, and complicated plumbing and electric systems in outbuildings." One of the biggest obstacles that Denise Molloy had "run into is just my lack of knowledge about how machines work . . . just the lack of skills. Growing up and stuff I didn't really develop those kinds of skills, so even like putting up the fence was just a struggle." Or as Diane Kaufmann put it, women have "never been encouraged to fix our cars or change our oil." Referring to the fact that mechanical abilities are not innate but learned, Jo Bauer said: "Men think they're genetically born with a means of finding the [oil] drain plug. But little girls, a lot of times they were never allowed or somehow had to have some reason to not be there, you know, crawling under the car with dad." To help address the perceived need for mechanical training, steering committee members organized conference workshops on such topics as "selecting and maintaining tools and machinery" and "farm carpentry."

Network members also discussed certain characteristics that tend to be associated with women and that many women have been socialized to think are less desirable when it comes to farming, most notably the perception that women have less physical strength and are smaller in size than men. A member explained that one of the advantages of the Women's Network is that you get "a woman's perspective" on ways to do certain tasks: "Men are physically stronger than women, no two ways about it, unless you're a very, very strong, large woman. So there are different techniques, for digging a post hole, as an example, that a woman can tell you, because she would do it differently than a man." An example of such an exchange occurred during one network discussion when a member commented about being small and feeling unable to do certain things physically. But another farmer, Sally McMahon, quickly pointed out that she thought her small size could be an "advantage." Disbelieving, several people asked in unison how that could be the case. Sally, a five-foot-tall dairy farmer, described how her smaller build means that she gets her cows to do things by "coaxing" rather than by "bullying" them as her larger male counterparts tend to do. And she told the story of a time that a male neighbor visited when Sally was trying to get a young cow used to its first milking. Sally leaned into the cow to calm the animal; then her neighbor offered to "tail" the cow so that its hind legs would be immobilized. She declined the offer because even though she knew such bullying would work, it would not work for *her* as a permanent solution. Moreover, she had learned that her method causes less stress to the cow and therefore more milk is produced. Her sharing of her personal experience with others in the Women's Network was an example of how one member concretely illustrated to others that there are alternative ways to think about the assumption that women's characteristics are negative. Those characteristics may even be advantageous.

In addition, network members recognized and tried to help one another overcome the invisibility of women's knowledge of and productive roles in agriculture, a problem deeply embedded in rural communities. Such devaluing can profoundly affect the course of a woman's life, as became powerfully clear at one network event. When it was Janet Richardson's turn to introduce herself to the group at the beginning of an event, she began crying and explained that her husband was asking for a divorce. The Richardsons were prominent participants in the sustainable agriculture movement and were well known, although she had never been to the Women's Network before. When Janet said she had dedicated twenty years to their farm, the group reassured her that they knew that she was an integral part of building one of the most celebrated alternative farms in the region. When Janet reported that her lawyer said that the laws made it much easier for her husband to buy her out than vice versa, the group expressed outrage. When Janet said she felt limited because she had only a high school education, the group responded by pointing out that she had twenty years of experience and knowledge about farming, which "counts more than any degrees," as one woman put it. The other women encouraged Janet to try to keep the farm if she wanted to do so and suggested she could take on apprentices and teach them what she knew. The discussion went on in this way for over half an hour.

I had seen Janet only a few days before this interaction at an event of another network made up of both men and women, and she did not make a public statement about her situation there, making it all the more striking that she could convey her feelings in a women-only space upon her first encounter with the group. Even for the Women's Network, however, the depth of emotion expressed was unusual. But Carrie Mann later used it as an example to illustrate a more general point about the way the group functioned:

I think not everything we discuss is of a technical or practical nature. It's more of a compassion, nurturing, empathy sort of level. . . . It's sort of like we're recognizing that there's more to what we're doing than the fact that we're just milking sheep or hoeing basil. . . . Remember Janet Richardson crying? I've never heard from her again. I don't know her, but she felt like it was safe to come to these women who she'd never met before and totally let loose her concern about her farm on a like totally what I felt was womanly level. Like she acknowledged who she was as a professional farmer and her doubt about that and just like let loose her feelings. She would never, I don't mean to make such a broad assumption, but I would say it would have been unusual for her to go to a male field day and respond like that.

So I guess that's sort of like a strong example to me of the crossover potential of a network of women farmers. Like there's a human element introduced, an experience introduced, that I just don't believe that men generally are relating on in their professional context. And I don't think all women would have felt comfortable doing what she did, but it was totally OK there that she did it. . . . That is one of the reasons why it's unique as a woman's experience in farming versus just an experience, a human experience in farming.

The discussion of Janet's situation focused not only on members' recognition that gender inequity resulted in Janet's work being devalued, but they also tried to *revalue* it for her and perhaps for themselves as well.

Another, more general and less unusual way that members of the network generated and shared an understanding of women's agricultural capacities was by being "role models" for one another. That is, in learning about the skills and achievements of other women who have struggled with and often overcome obstacles related to existing gender relations in agriculture, members seemed to obtain a crucial piece of knowledge: they too might be able to succeed as women farmers. Indeed, the members themselves identified the role model function of the network as particularly valuable. For instance, from Diane Kaufmann's perspective, one of the biggest obstacles that women farmers face is "things we put on ourselves or that society puts on us, not believing that we'd be serious about it or we could actually do

it. . . . Probably the biggest disadvantage is not being taken seriously."
Accordingly, upon founding the group Diane hoped the network would provide a space where women would take each other seriously and could learn
from others who have been successful. "The major thing is to find the role
models out there," she said. "You hardly ever hear about them. If one
woman is doing it, then I can do it. It's a really empowering kind of thing."
Similarly, for Denise Molloy the value of the network was "not just friend-
ships, but it's also the role models and, well, peers that you can kind of connect with from the production aspect of it too."

One principal way in which this role model function seemed to manifest
itself in the network was the considerable amount of time members devoted
to the introductions that customarily began each meeting or conference.
This practice began with the day-long conference in the spring of 1993. Diane and Denise invited the sixteen participants to "share your stories . . .
where you are from, history in farming, what you are doing now, your
dream for the future, any words of wisdom or helpful hints you've learned
along the way, and what you would want from this group." These introductions took two hours as women addressed some of these questions and were
often peppered with further questions and comments from others. Subsequent agendas always included sufficient time for this sharing because of the
strong appreciation that members expressed for hearing each other's stories.
Through such sharing, members seemed to teach one another by example.
That is, they articulated their commitments to their pursuits and displayed
their competence for others to see. In this way, these women farmers sur-
reptitiously exchanged the knowledge that, as Jo Bauer put it succinctly,
"you can do it."

Technical Information and Marketing Strategies
During the first several meetings of the network, the exchange of information about technical aspects of production and about marketing strategies
was overshadowed by the excitement generated from the very idea of forming a network for women, by the exchange of personal experiences related
to gender, and, as we shall see later in this chapter, by a strong emphasis on
the supportive function of the network. Exchanging technical information
was not only a lower priority, but there were also obstacles to overcome because of the diversity in members' agricultural practices and levels of expertise. Network members identified these needs and challenges early in the
group's development. But by the end of the period of this study, it was clear
that incorporating more of the technical aspects of alternative farming into

their network events had gained a higher priority. This was not surprising because these women farmers did not separate their practical commitments to agricultural production and marketing from their identities as women.

Steering committee members responded to interest in incorporating more technical discussions into their events by pursuing two strategies. First, they offered sessions on nonproduction aspects applicable to a wide variety of farm enterprises. Second, to facilitate knowledge exchange based on particular interests, they broke the larger group down into several smaller workshops to discuss specific topics. Sponsoring weekend-long conferences allowed more time for network events. Therefore, the group was able to offer a variety of workshops and still preserve enough time for the indispensable sharing of personal experiences through lengthy introductions.

Topics of technical interest that transcended the differences in the members' operations included subjects such as preparing a business plan, learning about government services available to farmers, and using soil surveys. But by far the most recurring topic in this category was marketing. Members shared their personal knowledge of marketing, and such exchanges reflected the fact that many women were interested in building locally oriented food systems and minimizing the distance between consumer and farmer that has developed in conventional agriculture. Typically, network members sold their produce directly to consumers or through a variety of unconventional avenues such as farmers' markets, farm stands, natural food stores, consumer cooperatives, and craft fairs. They discussed their experiences with these marketing strategies through the newsletter, in workshops, and during informal conversations.

One marketing strategy discussed in both the newsletter and workshops was the increasingly popular community-supported agriculture model. In a workshop on the topic, a member described how she and her partner gave consumer "shareholders" an opportunity to connect with food production and support a small farm through their CSA. She explained that their 130 members pay for a share at the beginning of the season, and in return a certain amount of locally grown organic produce is delivered to a drop-off site each week. The workshop covered other aspects of CSA's in detail, including incorporating educational programs into the farm's activities, budgeting for the farm's expenses, and making CSA programs more accessible to low-income people. And the workshop leader openly shared the challenges that CSA presented, such as the guilt shareholders feel when they are unable to use all the food provided or the fact that in the previous year the farmers made only half the personal income they had projected.

At another workshop, titled "Wool: The Second Income," Freddie

Moths described her own approach to marketing and stressed: "You have to worry about your reputation when you sell things . . . so never cheat your customer and never promise you can produce something in a given time unless you are certain because that can ruin you in a hurry. And if there is a flaw, be up front about it and give a discount." Freddie said she had come to learn that the spinners who buy from her "want to know about this sheep that this wool came from, that they're going to make this sweater from." To illustrate this point, she told a story that had really "struck" her:

One time a guy from Madison, there at the university, gave a speech and said show me a flock where every sheep is named, and I will show you a flock where they are losing money. And I thought, "No, sir." And I never forgot that. . . . My sheep are all named, I do make a profit, and they are named for a very good reason. My customers like to know the name of the sheep. They call up the next year and ask about that sheep. You have to consider who your customer is. . . . The point he was putting out when he said that was if you've got them all as pets, you are not going to evaluate them the right way. You are not going to cull as close as you should to make a flock pay. Some cases you could say that that's true, but not in my case.

Network members not only exchanged marketing information in workshops intended for such discussions, but they also shared ideas in a more informal way. For example, both Carrie and Diane marketed a processed food from their farm produce, pesto and sheep's cheese, respectively. During a break between sessions, Carrie, who was further along in her marketing of pesto, gave Diane the names of places and contacts that might be interested in carrying her sheep's cheese. They discussed the advantages and disadvantages of selling to chains of stores and the difficulties of having enough product to fill a market. While Diane described some of her "fears" about marketing, Carrie responded by explaining some of the ways she had gotten over her own hesitation about approaching people and her apprehension about buyers' possible rejection of the product.

Another major topic the network tried to address was members' perceived need for mechanical training. They did this through newsletter articles and conference workshops focusing on such topics as fencing, carpentry, and electricity. For example, sitting among a circle of women, Jo Bauer introduced her workshop, "Selecting and Maintaining Tools and Machinery," with general comments about safety and the importance of keeping tools clean and organized. Jo reached her large hands into a five-gallon bucket full of tools and parts from her tractor. Pulling them out and passing them around for people to see and hold, she shared her observations about choosing, using, and maintaining them. After a while, the group discussed different aspects a person might consider when buying a tractor, from figuring out how much horsepower is needed to finding parts for older machines.

While some workshops focused on a particular skill, others were more general and designed to give an overview of all or part of a member's entire operation. Sometimes referred to as a "farm case study," these workshops covered an array of topics that reflected the diversity of operations within the network, which can be roughly divided between those devoted to raising plant crops and those devoted to raising livestock. For plant crops, production-related topics included gardening for market, growing organic grain, and saving seeds. At one farm case study workshop, Faye Jones described her organic cut-flower operation. In a lively and detailed conversation, Faye covered her operation thoroughly, and members asked a lot of questions about different aspects of the enterprise, including such details as how she chooses the annual and perennial flower varieties she grows; how she prevents diseases; how she selects her favorite tools; how she harvests, bundles, and transports the flowers; how she developed a marketing relationship with several consumer cooperatives; how she built a low-cost, walk-in cooler to store the flowers between harvest and market; and how she keeps one bouquet for her own pleasure. Those women raising livestock – cattle, sheep, pigs, chickens – exchanged knowledge about rotational grazing, fencing systems, and animal health. For instance, one member led a workshop about her organic dairy operation, placing particular emphasis on using homeopathic remedies to treat illnesses in livestock. In addition to describing the principles of homeopathic remedies in detail, she discussed how she put those principles into practice, for example, identifying the appropriate remedy and administering the remedy to the animal. Rather than always relying on a veterinarian, she explained, "When you are an organic farmer, you have to rely on your skills more than outside resources. . . . Traditional veterinary medicine is expensive and doesn't treat the whole animal."

In their sharing of technical knowledge at network events, members often drew on their own personal, local knowledge and shared ideas about a wide variety of alternative farming practices and marketing strategies. As one member said: "Most of the women in the sustainable group tend to be very practical and give me very practical ways to help me overcome obstacles." Similarly, Wendy Everham appreciated that in the network, "I get information. I get good solid information. . . . I buy far less books than I ever have because now I know where I can go for information. And I'd rather trust talking to somebody who's done it, or who's doing it, and who's playing around with it than I would reading a book." Diane Kaufmann also compared learning from other farmers to learning from books: "You can get the theory from the books, but it's seeing it from the field days or hearing it and being able to ask questions about it that really helps you to understand it yourself and be able to bring it back to your farm and do it." Denise Molloy described the local knowledge she learns from others at the network:

It's on a level of, well, what works and what doesn't work. . . . There's always somebody that's farther along than you are, or that has tried something that you're thinking of trying, or has ideas about things, or just doing something different that's sort of like, "Oh, gee, I never thought about that." . . . It's mentally stimulating, all the different ideas and stuff. . . . It's experiential information I have found to be really important. And that's not the kind of thing you can always get out of books. Or if you read articles about it, you're not always getting both sides of the story.

Even though network members were engaged in very different enterprises, they appreciated the opportunity to hear what other women were doing on their farms, and this exposed them to new ideas that they may not have previously considered incorporating into their operations. For instance, Jo Bauer said she was thinking of adding a "u-cut flower" operation to her market garden after hearing Faye Jones describe her cut-flower enterprise. In addition, the diversity of farming enterprises within the network prompted Diane Kaufmann to observe: "The thing that diversity does is it doesn't create any kind of hierarchy. There's no best grazer, there's no best flower grower because we're all so diverse. There's experts in all areas. We're kind of equals. . . . And we're not competing with each other. . . . Each person who's coming in there is the expert in what they are doing." In these and other ways, members acknowledged that a great deal of practical, technical information was exchanged in the network and that such information was often valuable because it was based on personal experience.

Some members were more moderate in their perceptions of the extent to which they found the network to be a valuable source of practical information about alternative farming. According to several members, such knowledge exchange was less useful for those women who had relatively more farming experience than for those who were very new to farming. Jo Bauer considered the information available at the Women's Network to be "more basic" than other sources of farming information, but it was not necessarily a negative aspect of the network:

You also have to realize I've been doing this for ten years so a lot of it is basic. . . . But even if some of the information I know, you also forget things, and so it doesn't matter how basic it is. I realize that I'm learning things that I've either forgotten or just totally, you know, never even caught. And a lot of the conventional, or the other meetings I've been to, is they never discuss basics. You jump in there with both feet and then you either start swimming or you sink. Or you find some, some nice person, some nice guy that you can ask some really dumb questions of. And there's a lot of them out there, but you kind of have to go up to them apologetically and say, "I really don't know anything about this, could you maybe?" And you feel kind of like you end up putting yourself down.

Faye Jones explained what she got out of the network: "It's really less technical knowledge only because I'm more advanced in what I'm doing. I mean I always learn things. I think I share a lot of technical and growing information with other women." Or as Carrie Mann explained: "On the gross scale, I don't feel like I'm learning that much technical information that's applicable to my operation. . . . It's sort of the more subtle stuff that comes out – how people view what they do and how they're doing it – that feels helpful to me. The nuances of how something is done." Carrie went on to explain the "nuances" by referring to an instance when a member had flavored sheep's cheese in several different ways and asked others to evaluate how the different flavors tasted:

Just an example is these people were experimenting with their cheese and flavoring the cheese, and I was just really interested in that because that's a processed food, and how they were talking about it. . . . There's nothing really technical there for me to learn, but the whole process was exceptionally interesting to me. So as far as the information goes, at this point I feel like I've offered more than I've given in terms of like technical experience, and it tends to be more about marketing than farming, what I've had to share.

Jean Schanen thought that the network was not necessarily "an efficient method of delivery of hard information . . . that's not the main value." And she was skeptical of what she heard at the network: "Let's face it, you go to a meeting and you ask a question. If somebody answers the question, you don't know what's the authority for their information, and you don't know how complete it is, and you don't know what other information is out there. And so maybe it's fine and wonderful and maybe it isn't." Jean did, however, think that the network was a good source of information for "a specific contact, like here's a person who has this input. That's not something you're going to read in books. But that kind of information is extremely important . . . and I don't want to downplay that, but it's a very different kind of information as far as learning how to do sustainable agriculture."

The exchange of technical information was an essential function of the

Women's Network. If not, these women might have gone to other organizations for rural women. But network members consciously rejected those groups precisely because they ignored the technical aspects. This was clear, for example, at the first meeting of the Women's Network that Jo Bauer attended. When introducing herself to the group, she said she had been reluctant to come to the meeting. Explaining further, she relayed a story of the time she had agreed to attend a meeting sponsored by the Homemakers Club because her neighbor had asked her to go. She explained that the event was a "bad experience" because she learned how to "tie scarves" when she would rather have learned about how to "fix carburetors." The group roared with laughter. Clearly, Jo wanted information about the technical side of farming. But according to Mary Neth (1995:138), the Homemaker's Clubs were organized initially by the extension service in 1914 and ultimately served to alter "definitions of what was appropriate work for farm women" by concentrating on women's work in the home rather than on agricultural production. As a result, the information that would have satisfied Jo Bauer's interests was not incorporated into the Homemaker's Club agenda. But Jo and others saw to it that such technical information was a part of the agenda of the Women's Network. When the network developed a mission statement, Jo referred back to her previous comment and argued successfully that the statement ought to include a clear reference to the exchange of technical information.

Although I have discussed gender-related issues and the technical aspects of alternative farming and marketing separately, in practice these dimensions of knowledge exchange were profoundly interconnected. For instance, references to gender-related issues were woven into the technical discussion on selecting and maintaining tools and machinery, such as Jo Bauer's warning to the other women in the workshop she led: "When you call places to find parts, they will tell you a lot because you are female and they think you'll take their word for it. Don't do it." Recalling that workshop, a member pointed out the value of learning such information about selecting machinery from another woman: "I know mechanics, but I didn't know what kind of tractor we should be looking for and things pertinent to this. So someone like her, a female, was very helpful because females have trouble going to the feed store, the implement dealership at first until people, the men usually, working there get to know them, and then you don't have the problems. But it's easier learning some basics from other females before going into the men's area." And Faye Jones commented in an almost offhand way while describing what she appreciated about the network: "It's more of the aspects of women sharing information with women about what works for them, and many of us are struggling with, you know, being mothers, or single parents, or just what kind of grease do you put on the U-joint." These and other examples illustrate that the personal, the social, and the technical were fused during the knowledge-exchange process.

Sharing Personal Knowledge for Inspiration

By beginning their mission statement with the words "to inspire women farmers," network members underscored a fundamental feature of the approach to knowledge exchange in this network. My observations and analysis indicate that the network provided an opportunity for women farmers to learn from one another that the obstacles they confronted – whether stemming from gender discrimination or from the need for technical information about farming – could be overcome. Thus despite social or technical constraints, they conveyed to one another a sense of their own agency which members described by using such terms as *energizing, empowering, affirming*, and *inspirational* to help explain what they got out of participation in the network. Diane Kaufmann, for example, felt that the network was about "affirming the work that one is doing. It's acknowledging and affirming that what these women are doing, all their varied things, is good, and that it's exciting they're doing it. And it gives me another vision of what I could do. . . . It's empowering because of the encouragement. . . . You know, you can do it. And it's good that you do it."

For many network members, such inspiration and affirmation meant that participation in the group helped to "build confidence" in a way that was not available for them in the male-dominated sphere of agriculture. For example, participation in the network for Carrie Mann provided an unusual opportunity because most of her other interactions as a farmer were with men. But in the Women's Network, "We're not having to sort of prove that we are that which we say we are. We're just being it, and no one is feeling judged or judging about that." Echoing a similar sentiment, Wendy Everham described what she thought members got out of the network: "I think it's an emotional thing. They're just giving off positive feedback. They are in a room with other women, and they don't have to make excuses about who they are and what they do and why they love what they do. That's already a given."

Indeed, because their identity as women farmers was "a given" in the network, Faye Jones characterized the first meeting she went to as so "powerful" for her that she had attended every meeting since then. Her consistent participation led to a "sense of empowerment" that she described in the following way:

It has been a really positive impact in my life because I was feeling isolated. I think I was unhappy, just getting over my second marriage and moving from the big farm [out of state] back here. But the source of inspiration, and, I don't know how to say it, motivation, I mean I kind of wondered if I was crazy to try to do what I was doing, but I kept doing it. After the first meeting [of the network] I knew that I was doing absolutely the right thing and that I didn't have to have any doubt about continuing with the farming and moving in that direction to solely support myself . . . and to feel confident and to feel like I had a sense of support in doing what I'm doing. Also I realized at those first few meetings that, wow, I do know a lot about what I'm doing and it isn't foolish at all. . . . I just had to participate enough to start to see, feel really confident and excited about what I'm doing. It's a catalyst for moving forward with and getting clear about your vision of your ideas and your farm and what you want.

The process of network members being role models for one another and sharing their stories through lengthy introductions was a key mechanism through which the network functioned to inspire women farmers. As Denise Molloy explained:

It's kind of solitary when you're out on your own place doing your own thing, and especially when you're doing it by yourself. And then you come together as a group, and . . . it's really fun sharing what I've gone through and feeling sort of the energy in the room. I love hearing what everybody else is doing, and I feel myself giving energy out to them. One of the most fun parts of the group is when we all get together and kind of go around and share what we've been doing. It's just the energy in the room is always really high, and people are just genuinely interested in what each other is doing. And so then I guess maybe that's where the inspiring part would come, it's like "Gee, I'm on the right track" or it kind of gives you the little extra push to keep going.

Similarly, Freddie Moths felt incapacitated before she developed her sheep and fiber enterprise. Through the workshop she led describing her wool operation and through her ability to answer questions that others raised spontaneously, she displayed her own knowledge and her capacity to overcome the profound obstacles she once confronted. In this way, she demonstrated to other network members what is possible. In turn, because other members placed value on her work, which she felt was devalued elsewhere, she proclaimed passionately during the mission statement discussion: "When I come here, I feel good about myself."

Another related characteristic of personal knowledge exchange in the network was that the expression of "feelings" and "emotions" was an integral part of many of the topics members discussed. Such expressions ranged from one woman's "pleasure" of keeping the first bouquet of flowers for her own home before taking the others to market, to another woman's "struggle" dealing with her husband's "resentment" about how much time she devotes to her farming endeavor. And referring to her discussions with Carrie Mann about marketing, Diane Kaufmann described how the sharing of feelings was a particularly useful component of her own learning process:

I wonder if it's . . . dealing more with the emotional parts of it in one sense. Thinking back to Carrie and just talking with her about what she's been learning about marketing, it's being able to express your fears or concerns. You know, "How do you approach these people?" or "What do you do if they say no?" It's dealing with kind of those emotional things, feeling things that are never part of the conversation at a grazing conference, for instance, or what you get from a book. I guess it's being able to share the feeling part of it in your learning process that for me has been the difference in being able to say, "I don't know if I can do that" or, you know, "It's just so neat to see you do that because it's so scary to me."

Just the freedom to be able to express things, and that contributes to your learning because when you verbalize those things, then sometimes they become a little less intimidating. . . . Going back to the conversation with Carrie. It's like someone has gone before, and OK I can do that because she did it. And yeah it's going to be scary. They might say no, whatever, but she survived it, she's still walking and talking, and OK, I can get past that too. It's not just technique. It's dealing with the internal stuff. That to me is the beauty of the Women's Network because women I think are a little bit more free to, there's no stigma in sharing those emotions and feelings that men I think would like to in a lot of cases, but it's not part of their agenda.

The integration of emotion, in this case fear, as part and parcel of a very practical discussion about marketing reflects an insight often articulated within the women's movement. That is, as Wainwright (1994:79) argued, one "challenge to conventional notions of expertise" from within the women's movement has been "the insistence by many feminists that emotion can be productively combined with reason in the extension of knowledge."

Identifying as a Woman Sustainable Farmer

When nearly thirty women gathered at an informational meeting about the Women's Network held at the annual Upper Midwest Organic Farming Conference in 1994, some of those present identified themselves as farm wives and deplored the ways that "women are involved but not recognized," as one participant said. Others, such as Carrie Mann, voiced a slightly different conception: "I'm not a farm wife. I'd like to see more women respect themselves as farmers." Participants' view of their roles varied, but they seemed to agree that women's work in agriculture has been unfairly obscured and devalued. By articulating the idea that women should respect themselves and be respected as farmers in their own right, the network was constructing alternative identities. Moreover, those identities were specifically located in the context of the sustainable agriculture movement. These ideas, values, and beliefs (i.e., ideologies) about feminism and about agricultural sustainability will be explored below.

Women's Consciousness and Feminism

Throughout the history of the United States, women have formed many of their own sex-segregated groups (Bernard 1981). Frequently, organizations like mothers' clubs and women's auxiliaries have asserted and defended traditional gender relations. In other instances, a radical sense of group consciousness has emerged from women-only spaces, as was the case in many consciousness-raising groups in the U.S. women's movement of the 1970s. In agriculture, where women's involvement in farm organizations and commodity groups often takes the form of participation through auxiliaries, questioning of traditional gender relations has been uncommon. Carolyn Sachs (1996:18) has observed: "Rural women's knowledge and experience only rarely results in feminist politics and activism. Due to the particular patriarchal relations that characterize social relations in rural areas, rural women . . . resist but seldom directly challenge patriarchal dominance."

In the case of the Women's Network, members explicitly refused to create an organization that defends traditional gender relations. For instance, they decided not to become an auxiliary to an existing sustainable farming network in their area because they feared being relegated to a "subgroup." And the organizers of this network distinguished what they wanted to create from the conventional women's farming organizations with which they were familiar. Specifically, they wanted to form an organization that took women seriously as productive actors in their own right and that addressed alternative farming concerns such as the value of direct marketing over commercial sales of commodities.

Notably, members of the Women's Network did not explicitly embrace a feminist label for the organization. More often than not, they seemed to want to avoid addressing the issue directly. In part, this avoidance resulted from the fact that only a few active members identified themselves as feminist, while many felt unclear about the meaning of feminism. In addition, active members did not go to the trouble of conveying fully to one another why they felt a network for women only was necessary because they seemed to share an intuitive sense about that need given their personal experiences and their desire for knowledge. Others not in their situation, however, often seemed to have difficulty recognizing or understanding that need, as became clear to me when I observed numerous instances of people questioning why the women wanted their own sustainable farming network. For example, an article titled "New Sustainable Farm Network for 'Women Only'" in a Wisconsin agricultural newspaper made much of the issue:

Diane Kaufmann concedes to being challenged more than once as to why women need their own farming network, and admits she's even wrestled with that issue. After all, these are women who are struggling to be recognized as farmers in their own right. If so, shouldn't they be content to rub shoulders with their male counterparts at traditional meetings and field days? "The answer I keep coming back to is that there's such a different feeling of camaraderie and support that we (the women who've been involved in forging this new network) are reluctant to give it up," she confides. (Fyksen 1994a)

Other network leaders and members also wrestled with ways to articulate why the network was meaningful to them, especially when they applied for funding from the Sustainable Agriculture Program and developed their mission statement. Despite their strong enthusiasm and apparently intuitive sense that the network was extremely important to them, members were not always conscious of or able to express exactly why it was meaningful, as Faye Jones explained: "It's something we don't have words for in our vocabulary yet." Frequently, members used the word *support* to describe what they got out of the network; however, what is of interest here is the extent to which members connected that need for support to their experiences with inequitable gender relations.

Like gender discrimination everywhere, gender relations in agriculture can be experienced by a woman without her recognizing it as such. In other words, the everyday world is organized by social relations that are not necessarily observable or completely understandable from within it. One way to gain a better understanding of the organization of social relations is through a deliberate process of sharing personal experiences and analyzing the recurrence of those experiences to find common causes within social arrangements. The consciousness-raising groups used by feminists in the 1970s as an organizing tool constitute an explicit example of such collective reflection on the political meaning of personal experiences (Evans and Boyte 1986; Ferree and Hess 1985).

Jean Schanen clearly interpreted the sharing of personal experiences of

gender relations within the Women's Network as akin to the feminist practice of consciousness raising. Jean identified strongly as a feminist; for example, at the first network event she attended, she introduced herself to the
group by saying: "I've been a feminist for a lot longer than I've been a
farmer. . . . I could immediately see the value of a Women's Network." In
her interview, she later clarified what she meant by the comment and compared the exchange of personal knowledge in the Women's Network to consciousness raising in the women's movement:

We all, as we go through our life's endeavors, we need to feel like we're really capable of doing what we're doing. And the male-dominated culture doesn't give us a
whole lot of that. That culture teaches us to doubt ourselves and to question and to
think that we probably can't do it and that we need somebody else to figure things
out for us. The big thing that I think the women's movement has offered in general
and has certainly been offered in the women's sustainable ag network is the very certain knowledge that we are not alone, that there are other people who are working on
the same kinds of issues and having the same kinds of experiences. And then if we
have a problem it's OK because it's part of the general experience. It's not because
we are weak and helpless. . . . That certainly is what I got out of the women's consciousness-raising groups of the seventies, and I see the same kind of commitment
going on here. . . . For the most part there's no physical support or very concrete
support provided [by the network]. It's much more spiritual. . . . And to me that
kind of inspiration means believing in yourself and trusting your own capabilities,
and deriving from that a feeling that it's just fine to go ahead with what I'm doing
because I can do it.

While Jean clearly identified as a feminist and associated the network
with feminist principles, not all members shared that perspective, as became clear during what was the most sustained discussion about feminism
that I observed at a network event. On the first night of the weekend-long
conference, Diane Kaufmann welcomed the participants and explained the
history of the network, stressing: "What we have found is that there is a
good feeling from being with women and wanting time for women to talk
with one another." She went on to say that "this is not a feminist organization because we are not trying to create lines between women and men."
The issue arose again the following day, when one network member led a
workshop describing her farming operation and introduced herself by explaining that she had started a newsletter for "rural feminists" from around
the country. After she explained the history of the newsletter, one woman
asked the workshop leader, "Could you give us a current definition of feminist?" She replied, "I identify as a feminist, and I think you are a feminist if

you think you are." Then Jean Schanen referred to Diane's comment from the previous night: "Diane said this is not a feminist organization, and I was shocked because I think it is feminist. Why did you say that?" Diane explained, "I just meant that whoever wants to come to the meetings should come and that I don't want to set up barriers."

This interchange was further clarified after the workshop, as a few women were sitting around a table talking and waiting for the network's business meeting to begin. When Jean joined the small group, Diane said she thought she needed to clarify what she had meant by her statement about feminism the previous night. In her view, the network should be open to anyone who wants to come to it, whether feminist or not, and she added that she does not know for herself what feminism means. Jean replied by explaining that her view of feminism came from the consciousness-raising groups of the 1970s where women "developed trust" and "were trustworthy," and she thought the network was similar to the consciousness-raising groups in this way. Several other women added that they too did not know what feminism means, and such confusion over terminology was compared to the way that people disagree on what is meant by the term *sustainable agriculture*. Later, when the business meeting began, Diane stood up in front of the group and stated: "Last night, I said this is not a feminist organization. . . . What I meant to say is that I want any woman interested in agriculture to feel welcome. It is not that feminists are not allowed, but if you are not a feminist you are welcome too." Faye Jones asked, "Did someone give you shit?" Diane replied, "No, just consciousness awareness raising."

In sum, the Women's Network as an organization represented a departure from many other women's groups in agriculture in that it affirmed women's productive roles and focused on alternative farming practices. Although some network members proudly identified as feminist, others were leery about adopting a feminist label for themselves and for the organization. This latter position carried over to the network as an organization, as indicated by the concern that a feminist label for the group would discourage the participation of women who were suspicious of feminism. As one woman claimed early in the network's development, "A lot of women don't come because they think it is going to be male bashing and to be a bunch of radical feminists." Although the network may not have identified itself as a feminist organization, there was room for such a consciousness to emerge as they shared their personal experiences with gender relations in agriculture. Unlike the consciousness-raising groups of the 1970s, however, the Women's Network did not involve a deliberate effort to characterize and understand the social processes underlying those everyday experiences. The extent to which that happened – and it did happen, as shown in the previous section – was unintentional and was not articulated as a purpose of the organization.

Meanings of Agricultural Sustainability

Members of the Women's Network identified strongly with the sustainable agriculture movement, often making positive statements of support, such as Denise Molloy's comment at one conference that "sustainable agriculture just feels right." In the course of network events and in the newsletter, members articulated a broad vision of agricultural sustainability, and they elaborated on their ideological commitments further in interviews with me. Wendy Everham referred to this broad view, explaining: "To me sustainable agriculture has a lot more dimensions than just the farming. I think it also has an economic, a political, a social, as well as environmental dimension to it." Similarly, Carrie Mann's goal was holistic, and for her sustainable agriculture meant "farming in such a way that all the elements involved are sustained and are nurtured by the process, by the very process. So the soil, the plant, the etheric level I guess that's around the plant, and the insects and so forth, and then the humans who do the labor as well." And Jean Schanen explained at one network event that she was trying with her market gardening project "to model various ways to engage in agriculture, ways that sustain our bodies and spirit and the Earth." The ideological frameworks that constituted these multidimensional visions of agricultural sustainability seemed to influence the content of the knowledge exchanged among network members, particularly those exchanges related to technical information and marketing strategies.

For many members of the Women's Network, a sustainable agriculture would be a food and agricultural system that could perpetuate itself. As Ann Hansen put it, sustainability means "being able to farm the same piece of land for thousands of years without becoming dependent on outside inputs. . . . You can keep on doing it. I think of what they do in China. They've been farming for forty centuries in China and that's basically without a chemical industry. That's what we should be aiming for here." Such an ability for a farming system to perpetuate itself means that "your farming practices put in more than they take out," as another member explained. Similarly, for Diane Kaufmann sustainability "means doing the positive things in your farming practices that will better the land or your animals rather than mining it, just taking away. I think it's looking for production practices that aren't dependent on buying the chemical inputs and trying to, I guess, work more closely with what you've got in your setting, with nature. . . . So sustainability would be leaving that ground and those animals better than when you got them, so that the next generation can farm and be productive and not have all the problems we're seeing now."

Embedded within members' conception that agriculture ought to be able to persist was a fundamental tenet of the sustainable agriculture movement more generally, that is, a commitment to reducing significantly or eliminating entirely the use of synthetic pesticides and fertilizers, as well as minimizing the use of fossil fuel energy. Most members reported that they did not use what many referred to as "poisons." Although Jo Bauer occasionally used synthetic fertilizers, she cited her long-held respect for Rachel Carson and asserted, "Spraying, that's definitely a no-no." Jean Schanen described sustainability as "an ideal that perhaps we never achieve. But it involves things like balanced energy budgets, which means consuming no more energy than is produced in the calories of food production. It involves minimizing inputs. . . . We're trying to use ingenuity instead of machines and purchased inputs to accomplish our growing task." For many network members, reducing or eliminating their use of agricultural chemicals and machinery was intricately tied to their conceptions of both environmental stewardship and farm profitability.

Another element of sustainability articulated by members relates to what Jo Bauer called "earth-friendly ethics." Members frequently made various references during network events to their goals of land "restoration," "soil regeneration," and "taking care of the land." For instance, Carrie Mann spoke of the critical importance of maintaining the soil: "Without nurturing the soil, the soil cannot produce. It's not sustained. . . . I think it's really, at this time, really fundamental that we look at the ground because it's the basis from which all things come. And if we don't take care of it, it's all over." Faye Jones felt a similar urgency about the state of contemporary agriculture. Here she refers to her own farming, as well as the advocacy work she does with alternative agricultural organizations:

There's my farmwork, but then there's the other stuff I do, and they're related. I feel very pulled to make a difference in the world, and it seems so incredibly clear to me that food is something we all need, we all have to deal with every day in order to survive. It's one of our few basic, real survival things we need to survive. Too many people take agriculture for granted. Agriculture is the number one source of nonpoint [source], groundwater pollution. It doesn't have to be. You know, there's chemicals in the groundwater. Most people don't realize how much our current conventional agriculture destroys the planet. Yet the very thing we need to survive is food; but what we're doing is creating our own death, if we don't make some changes. And it didn't take me long to see that you don't have to use the chemicals.

There's lots of very successful farmers not using chemicals. There's a big myth about "Oh well, you can't really farm organically," "Oh, it works for some people," "Oh, you can do it sometimes." And I'm trying to blow that out of the water. I basically want to change the way America farms.

Such concerns for the environment inspired members to adopt particular approaches to land management. For example, Denise Molloy chose to raise sheep because it was "a way of utilizing the land. It's a resource that is available to us, and it's our responsibility to take care of it. . . . We want it to be better after we're not here any more than when we got here, so that's why I clicked into the sustainable ag movement, because in my heart I believe in that philosophy. . . . It's a philosophy of farming in which all your decisions are based on "Is this good for your land?" or "How is this going to improve your land?" instead of saying "How much can I get out of it?"

Members of the network were concerned not only with the environmental impacts of their farming practices but also with the profitability of those endeavors. According to Lois Nerby, sustainable agriculture means being able to "produce without having to buy a lot of stuff." Jo Bauer explained further: "The more nutritious yield I can make with the least amount of inputs, of course, I'm going to be more profitable. . . . So a lot of the sustainable things are actually economic things too. They work kind of hand in hand, and I don't always say we're going to do this because it's ecologically right. It's kind of like, we're going to do this because we can't afford not to." Even as members tried to reduce their costs of production, many of them continued to struggle financially. Such challenges prompted Diane Kaufmann to state during the introductions at one conference that she was "trying to figure out if farming means taking a vow of poverty. I hope it doesn't." Carrie Mann responded to this comment by saying that sustainable agriculture must include "economic sustainability." To explain, Carrie said: "I feel like if many farmers cannot survive financially, that's not sustainable. That means that farming is not sustaining the human component, and again I have to say by its very nature, agriculture involves the human component. So I think that part of sustainable farming is doing it in such a way that it's economically viable. If growing basil wasn't providing for my basic needs, it wouldn't be sustaining me."

Many members coupled their concern for profitability with a commitment to reducing the social and spatial distance between producers and consumers. Wendy Everham put it, "When I think of a sustainable food system, I think in terms of a more regional or local level." Flower producer Faye Jones wanted her farm to contribute to building a locally based food system, as she noted when talking about her goal to add vegetable and fruit production into her farm:

I think ideally we need to get more to a society that really thinks and acts locally and isn't expecting strawberries in February from California. . . . We're going to need to have much more emphasis on local food systems. You can still have a wide variety of good food, but I think it's inevitable that we're going to have to start producing more of our own food around our major metropolitan areas. I don't even think it's an ideal, I think it's something that's just going to happen. The cost of gasoline to truck this stuff or fly this stuff and all that is starting to make it not as appealing. . . . I also think people care enough about their food and the environment to say, "Hey, I would rather buy from that woman down the road that I know than somebody in California" because they're starting to connect that agriculture can be one of the most devastating impacts on the environment. If you're willing to recycle, then you need to take that next step in realizing that where your food comes from and what you eat every day, many times a day, has a profound impact on the environment.

It's a really important point that most people don't have that connection. Or they may say it, but they don't take any action at all, and it doesn't mean always buying organic food. It's much bigger choices than that. It's being willing to say, "Well, I'm going to see what I can do to support my local farmers because I don't want it to be a ten-thousand-hog operation down the road instead of three dairy farms." I think that if people can start to see it in those terms, they can make more choices. . . . That vision plays into how it is my farm is going to relate to the community as far as supporting food for more of a local-based thing. My farm operation will very much play into that changing idea of where our food comes from and why we should care.

As members shared information on marketing, this emphasis on locality and on building relationships with consumers was evident in many network activities. For example, in a newsletter article about marketing tips, one member wrote of the importance of cultivating a core group of customers because "the business we are in is really that of relationships." And at a conference workshop, one member shared the mission statement of her CSA: "To provide farm-fresh seasonal food produced in a manner that protects and enriches the land and people. To be a community that links urban and rural life through shared work, play, education, and food." Similarly, Jean Schanen felt it was "intensely important" to promote "the empowerment involved when people have control over their own lives, where people can make real choices about what kind of food they're going to eat. I mean, food is just so central."

The idea that food is central was echoed by other members who thought that food was an important educational tool and that farmers played a role in promoting that education. As Wendy Everham expressed the idea: "I just want this to be a far more sustainable society in every way, not a throw-away society, not a society that thinks it can spray its way out of problems, to just be a little more thoughtful about what we do. And it seems to me that maybe food and food systems are one of the best ways of beginning that kind of dialogue because everybody eats." Diane Kaufmann felt that such a dialogue began when her customers came to her farm to pick up the broiler chickens and turkeys she raises on pasture, and even though direct marketing was not always easy, it had rewards for her:

I know how little people know about where their food comes from, and I know how important it is that farming and producing food is done sustainably. I guess I feel that kind of a mission in a sense, although I like to demonstrate it and not beat people over the head with it. . . . Connecting with people, having them come here, it boggles my mind that people have no understanding at all of where their food comes from. It's just kind of fun to jog their latent memories about that or just show little kids baby chicks or just to show them the reality of a farm too, that there is mud and manure on a farm and it isn't all little white picket fences. It can be a nuisance doing that when you deal with people who don't show up to pick up chickens or stuff. But I think overall the feedback you get from your grateful customers about your product – when they tell you that they've never had chicken that tasted this good or that lamb was the best we've ever had – I mean there is just a very intangible reward in that too that you don't get if you shipped your animals to the packer or something.

For a few members, building a local food system ought to be connected to the need to address the problem of hunger. Wendy Everham echoed other members when she said: "I'm also concerned that right now farmers' markets and CSAs are really middle class. We should broaden the concept [of sustainable agriculture] to be able to offer food to people who don't have the money to be able to afford it."

A Strong Support Network

When members of the Women's Network developed their mission statement, the facilitator sensed the strong enthusiasm among participants and asked, "Why does the network make you feel good?" One member replied, "Because it affirms you," and another added that "you don't feel odd here." That sense of affirmation and belonging could be explained, according to a third member, because "the whole term 'farmer' implies male, but the women in my neighborhood milk the cows. . . . What we are seeing is that people don't question the way it has been, the way that there are farmers and farm wives." Other members responded with laughter that seemed to acknowledge an unsettling and ironic truth. Many members of the Women's Network rejected, either directly or indirectly, the stereotypical role of women as farm wives rather than farmers. Moreover, from their personal experiences, these women knew that claiming the title of "farmer" for themselves crossed accepted gender boundaries in ways that were both personally and materially challenging to them. Accordingly, by creating a space for women in sustainable agriculture to exchange information and ideas, the network provided much-needed support to its members as they crossed those gender boundaries.

Unlike the grazers' network, whose members mentioned to me only in interviews the social support function of their network, members of the Women's Network repeatedly stressed the supportive aspect during conversations among members, in large-group interactions at network events, and when talking with the media. Thus when members formed their mission statement after more than a year of meeting with one another, it was hardly surprising to me that they agreed immediately on explicitly including the goal of building "a strong support network."

Many members openly referred to the "camaraderie," "good energy," and "support" that tended to permeate events, feelings that seemed to be derived from the opportunity to be with other women interested in farming in general and sustainable agriculture in particular. For example, Jean Schanen described her first impression of the group:

My first meeting that I recall was at the church. It was just a gathering then, I think there were only fourteen women. And it was primarily just farmers talking to each other, and I found it to be very exciting and uplifting and wonderful, and other people did too. That was, I think, the most common sentiment expressed. It was just merely being in the room with other women farmers and having that kind of support from each other; it was a real high-flying experience.

These sentiments were commonly expressed. Often characterizing the network as a "safe haven," Diane Kaufmann referred to the ways the group instilled "the sense of support that you're not strange for wanting to be a farmer, that other women feel the same way." And Freddie Moths felt the network offered her "moral support" because members shared a "common interest" in farming:

What I get out of [the network] a lot of times is listening to the other people and talking. Not necessarily that I do the same things as they do, but it's sort of like moral support, I think that's what I'd say, mostly moral support. I enjoy it 'cause I feel right at home. . . . With the women that are there I don't feel out of place because we all kind of have a common interest in farming, one way or the other. Some are raising this, some are raising that, but we still have this common interest.

Moreover, because women identifying themselves as sustainable farmers deviated from the norm in rural Wisconsin, opportunities for support were not available elsewhere, as Jo Bauer explained:

There aren't too many places you can go in the area where you're looked at as a female that has any and gets any level of respect. . . . I can go to a meeting with [my husband], and everybody will say to him, "Well, how are things doing?" And it's kind of like, well, I don't exist it seems. . . . This [network] is a group of people who I know full well that every woman in that group has calluses, gets her hands dirty, and is an integral part of the operation. . . . And it's just something that's totally lacking out in the male-dominated world of agriculture.

Although many of the struggles network members encountered stemmed from gender discrimination, for some members the need for support also emerged from the difficulties they encountered when dealing with class discrimination and prejudice against farming and rural life. For instance, Carrie Mann reported to the group that her mother, apparently unimpressed by her daughter's chosen vocation, "always asks when I'm going to get a real job." And Wendy Everham, who raised a large vegetable garden for some twenty years in an urban area before she moved to Wisconsin, never before felt comfortable sharing her interest with family and friends, who, in her view, looked down on gardening as "lower class." According to Wendy, gardening had been "an aspect of my life that I found joyous, but I never talked about, because I didn't have anybody to talk about it with." In the Women's Network, she found that

even though we're not doing the same kind of farming or growing at all, these are women who could talk about manure, and you could talk about fertilizer at lunch, and nobody says, "This is not polite conversation." . . . Just to be in a room with women who are doing similar things that I love to do, and that there are people I can talk to about these things and that's considered serious conversation, it's just an accepting kind of thing. And it's just a kick for me that there are that many people in the world who don't worry about what expectations other people had of them, that they just went ahead and did what they wanted to do. . . . You can just be comfortable talking about things that maybe other people would think you were weird.

Likewise, Ann Hansen found in the network others who are "as crazy as I am" because they have chosen to farm:

I went to that first meeting thinking I was really really a weird person. Here I was city-educated. I have a college degree. I just have an excellent background. I was in the business world for quite a while and doing fine. And just something was wrong with me because I wanted to live in the country and get dirty. And you walk into that meeting, and there's twenty other people, and they're all doing the same things. They're all very bright. They're all wondering what's wrong with them, at least that was my impression, because we want to farm. We all sit around and talk, and you start to realize this is a good idea. It is a good idea to want to live in the country and to bring up your kids on a farm and to be outside all day and, you know, mess around with animals and plants and machinery. You feel like you're not alone. You feel like you have people to turn to when you have problems, and that's support.

While these sentiments seemed to be expressed most frequently by members relatively new to farming, Freddie Moths, who was raised on a farm, articulated a similar idea:

I guess sometimes, when I get in company with other women that are, oh what would you say, all dressed up, makeup, and working out [i.e., at a job outside the home or farm] and strictly city, I feel kind of out of place. I feel at home there [in the network]. I feel like I fit in. . . . I come from a family where there's five of us. And the other four, even though we were all born on the farm originally and we lived in the country, they have all gone citified, and they have their jobs. They kind of look at me kind of in amusement, 'Freddie with her sheep on the farm.' They don't take it, me, or my sheep very seriously, as a serious job or anything like this. I mean I'm just playing with them sheep, you know, and they don't really see it. Now if I had a serious nine-to-five job, they'd take me serious.

Another aspect of the network that members found supportive dealt with their perceptions of differences in the way that women relate to one another in comparison to the way that people relate in groups that include both men and women. The Women's Network provided what they called a "comfortable atmosphere" where there were "no stupid questions" and no "competition" among the members. It was these impressions that led members to refer to a "different feeling" in the Women's Network and to indicate that in other settings, particularly male-dominated or mixed-gendered groups, these feelings do not exist. For example, when Jo Bauer attended her first network meeting, she referred to her perception that other farming meetings are "95 percent men" and that she looked forward to the "open sharing" among women. This openness was something she found lacking in male-dominated groups: "To be honest, they lie through their teeth. You ask them how it's going, and they are standing in manure up to their knees." Similarly, Ann Hansen said: "I think women communicate differently from men. . . . The bullshit gets this deep at men's conferences because they're all trying to be the biggest dog on the block or something like that, and they lie. . . . I don't feel that goes on at the Women's Network. . . . I think women tend to be better listeners than men. . . . I think women are much more cooperative rather than hierarchical."

In addition to members' perception that women are less competitive than men, they also felt that a key difference between the Women's Network and other groups was the willingness of the women to share and discuss their "feelings" and "emotions." For instance, Diane Kaufmann observed:

There's definitely that sense of affirmation and energy, I guess, that comes from being able to be with a group of like-minded women who probably, what do I say? It's kind of a male/female thing in a sense. It's just a totally different kind of thing when you're with a group of women and maybe the feminine qualities that come out. There's not a sense of competition, there's a sense of nurturing, caring, wanting to see you succeed, feeling, expressing, maybe just being more free to express those kind of thoughts to one another, that those are important things. And just the excitement and the energy that is in that group, and it happens every time which is kind of an amazing thing, even though this last time [the network got together] the balance was really three-quarters [of the women there] were totally new people, and still you got that same experience happening again. So it's like you know that when that situation is provided or can happen, that it's meeting a need that a lot of people have felt.

Similarly, Denise Molloy compared participating in the Women's Network with another network she had been involved in that was primarily, but not exclusively, made up of men:

At the women's group meeting we talk about things, and we can come at it from any one of a number of different angles, and everybody is comfortable with every angle. It's like you can get into a discussion about the nuts and bolts of things, or what type of fencing to order, or where to order it, or the real kinds of things that the [mixed-gender network] gets into too, and everybody feels comfortable talking about. But then you can also get into a discussion on balancing the farm with your family, how you feel you're pulled in a bunch of directions, or how do you manage this part, and everybody clicks in on that one. Whereas in the [mixed-gender network] men would say nothing, it's just a whole issue that they don't feel comfortable with. . . . I really think women in general, not everybody, but in general women are more comfortable talking about more personal issues than men are. Even though men might be willing to discuss it, there isn't the same feeling in the room. The feeling in the room at the women's group, it's just much more supportive. It sounds so vague. It's an atmosphere of sharing that's really promoted, and it feels comfortable. . . . I guess it's just the atmosphere where you feel comfortable saying whatever you need to say, and you feel like you'll be accepted for what you're saying, that people will listen to what you have to say, and comment on it, or offer information if they have the information. It's not a competitive atmosphere, it's a cooperative atmosphere.

In the course of network events, members not only shared their "feelings" but also their "mistakes." Like the grazers who appreciated the opportunity to hear what others had learned from trial and error, so too members of the Women's Network appreciated participants' willingness to express their "foibles," as Carrie Mann described it. But what distinguished the Women's Network from the grazing network in this regard was that these women felt more "comfortable" sharing their struggles in a women-only space, a characteristic that members associated with the supportive function of the network. Jo Bauer appreciated the way that members freely shared their mistakes without judgment by others:

Part of the problem that I find with conventional agriculture is nobody ever wants to admit their failures or their disasters or the things that didn't quite go well. And that's what I like about the Women's Network is that it's easy to say, you know, this didn't work too well. It's like nobody ever judges you. . . . It's just real positive. There's more to agriculture than "How's the corn growing?" with the Women's Network. I mean you want to know how to make the corn grow, and you want the corn to grow well, but there seems to be, I hate to use that term, but maybe nurturing, because it's such a feminine-type term. But I think that there's a nurturing sense under everything.

Faye Jones described it this way: "When us women are together, it's a much different interaction than it is in other meetings that I attend where there's mixed sexes because no question is foolish, no thought or feeling is. You know that no matter what you say or do, it's okay, even if it's goofy or off-base. And there's a certain comfort in that." And Carrie Mann summed up her experience with the group: "I feel like what I know is respected, and what I have to learn I respect, and it's all done in like sort of a funny and loving way."

That the network seemed to provide a uniquely comfortable setting for members to discuss their successes, failures, and feelings was intricately related to the perception that they had shared very similar experiences in an agricultural system that tends to be male-dominated. Of course, there were numerous differences among members in marital status, age, socioeconomic status, and farming experience. While acknowledging the existence of these differences, members emphasized that one of the reasons they found the network to be supportive was that the women had certain common understandings that emerged from similar experiences in a gendered society. For instance, one member said she particularly enjoyed the opportunity for "sharing experiences that are typical to women," which she explained further:

Men don't have problems going into the feed store and getting feed because they're dealing with men; or if they do, they have problems dealing with anybody. Men don't have problems going to the implement dealership and trying to get their respect initially. Even though a woman may have all the information in the world, and you may have torn apart and overhauled three tractors, they don't know that and they treat you that way. So, women have all had similar experiences, and you can talk about, well, how you overcame them or what you did in this situation. . . . It's hard to deal with men when they're not used to dealing with women.

Likewise, Diane Kaufmann felt that members of the network "have shared similar experiences or have gone past those experiences and can kind of lead the way," and, as a result, she felt there was a sense of "empathy" in the group:

Because you know that the gal sitting next to you has probably gone through many of the same experiences – trying to balance work, kids, husbands, if they're in that situation, or especially women who didn't grow up on a farm that are now involved in it, learning how to deal with the feed salesmen or whatever. It's like you know you've all shared probably some very similar paths and experiences that maybe some men have, but probably the majority haven't. It's been totally different for them. It's like the women in the network have shared, I guess, it's you've shared a common experience. There's been a few women [in the network] who are more traditional farm women who grew up on dairy farms and are now running their own farms. They certainly probably didn't go through many of the same experiences in learning how to farm as we did, the rest of us did [i.e., those who did not grow up on a farm]. But they are certainly balancing all those other things in their lives and making it work. Just an empathy.

For similar reasons, Denise Molloy found that "in the women's group you can talk about feeling like you're the odd person out at the feed mill or whatever and everybody knows exactly what you're talking about. Or like I need to make arrangements with neighbors about getting my hay done, and some of the struggles just being a woman trying to do that stuff. Everybody has been there. So I guess that's what I like about it."

Network members felt that the support they derived from the group would ultimately contribute to reducing the obstacles that women farmers confronted. Nearly all members held an incremental view of social change, arguing that women had to "prove" themselves to others. As Jo Bauer put it:

Part of it is you have to stick up for yourself. You have to sometimes point out to people what you have done because they sometimes don't see what you've done. And you have to be aggressive enough to do that. Otherwise, I think what it's going to come down to though is we're operating in a more of a male-dominated world than probably most of us would like to. We just have to prove ourselves, and the results are going to have to speak for themselves. . . . I think you have to get out there, and the people that frown upon you have to know what your accomplishments and achievements are. Sometimes you have to bend over backwards to accommodate them, but sometimes you also have to kick them in the butt and say, "Listen, this is what I've done, and I'm proud of it." I think a woman's network, though, is an integral part of that because you can't do that without some support behind you. And if it involves having a field day for the Women's Network and inviting everybody in the state to come, then that's the way to do it too. . . . I think women are such a unique part of agriculture that they need to stand out there very proudly and say it.

Likewise, for Diane Kaufmann one way some of the obstacles that women confront will be overcome is "by having successful women out there running farms, being financially successful. . . . The more women that we see out there farming, the more you have to take it seriously." Her vision of where the Women's Network should go in the future was connected to that idea: "What I envision is helping make each other successful. I don't see it as a political platform. I think there are other areas that can be done. To me the most important thing it could do is just help make each woman successful in what she's doing, helping provide the tools . . . helping each other to realize our goals."

Members of the Women's Network created and exchanged personal knowledge in order to enhance women's capacities to practice sustainable agriculture. In their network exchanges, members drew on their personal knowledge derived from their social location in a society that has consistently undervalued women's contributions to agriculture and from their local knowledge generated during their direct experiences with alternative farming techniques and marketing strategies. Although not all members identified as feminist, their exchanges were reminiscent of the consciousness-raising groups in the women's movement several decades ago, particularly the ways their exchanges in the network conveyed a sense of personal agency and inspired one another to continue their efforts to realize their goals despite personal and material challenges. Network members also articulated strong commitments to a broad vision of sustainable agriculture. Perhaps most important, from the perspective of members themselves the network provided a supportive and comfortable atmosphere in which their knowledge and identities were respected, an opportunity not available to them in the male-dominated field of agriculture.

SIX

Trading Ideas and Transforming Agriculture

Many of the family-owned and operated farms and ranches that persist despite today's industrialized, globalized agriculture continue the tradition of helping their neighbors when there is an illness in the family or when there are periods of heavy work, such as bringing in the harvest or moving cattle from summer to winter pasture. In fact, "to neighbor" is still a verb in many parts of the rural United States. But something more than just mutual aid neighboring is seen in the activities of the sustainable farming networks studied here, something akin to what the rural sociologist Carl C. Taylor advocated more than fifty years ago. In a pamphlet he wrote for the U.S. Farm Security Administration, Taylor (1941; see also Christie 1996) encouraged farmers to visit with their neighbors and systematically "trade ideas" for educational purposes. Taylor proposed that by sharing their experiences and ideas with one another, neighboring farmers could collectively refine their "common sense" and make it most relevant to the place where those ideas were put into practice. In a statement that seemed to anticipate the importance of the exchanges that occurred within the sustainable farming networks explored in this study, Taylor wrote:

Trading ideas is different from trading any other commodity in the world. This is illustrated by the old statement, "I have a dollar and you have a dollar. You give me your dollar and I give you my dollar and we each still have just one dollar. But I have an idea and you have an idea. When we have traded ideas, we each have two ideas." Trading ideas goes even further than this, for sometimes one or the other of us, or both of us, will get a completely new idea just because we talked things over. (1941:1–2)

The essential characteristic of the farmer networks studied here is not simply that members exchanged ideas but that they exchanged ideas they generated themselves.

During the many years over which the formal institutions of agricultural science have neglected alternative agriculture and delegitimized farmer-generated knowledge, farmers in the sustainable agriculture movement have produced some of the knowledge they needed. In addition, farmers and other advocates of sustainable agriculture have created mechanisms through which farmers' local knowledge can be shared and have thereby established an alternative knowledge system outside of the formal institutions of agricultural research and extension. This alternative knowledge system now has local expressions in nearly every state in the country and includes a range of approaches to the production and dissemination of knowledge – from research institutes made up principally of defectors from conventional science to local clubs and networks consisting primarily of farmers, rural advocates, and those maverick agricultural scientists and government agents willing to work with them. This resurrection of farmer-generated knowledge constitutes one strategy for responding to the ways that agricultural research policy and the structure of agriscientific institutions have contributed to the industrialization, globalization, and corporate control of agriculture that the sustainable agriculture movement has found to be socially, economically, and environmentally problematic.

The research described here has focused on how the creation and exchange of local, personal knowledge constitutes a principal social movement activity in two particular embodiments of this alternative knowledge system, the Ocooch Grazers Network and the Wisconsin Women's Sustainable Farming Network. Although the rich potential of local knowledge for sustainable agriculture has been recognized previously as a theoretical concept, the two networks studied here provided concrete settings in which to consider how the creation and exchange of local knowledge function as a practical foundation of this social movement. Expanding on current interpretations that emphasize the deeply personal and tacit character of local knowledge production in agriculture, my analysis of the two networks suggests that local knowledge was being socialized in these settings, albeit sometimes in different ways and for different purposes in each group. And when examined in light of recent advances in social movement theory, the study of these two networks helps to better articulate the importance of farmers' production and exchange of knowledge as a local-level approach to social transformation in agriculture.

Exchanging Knowledge, Building Community

The two cases that constituted the focus of this research offer instructive comparisons especially with respect to their organizational structure, their production of local knowledge, the exchange of local knowledge among network members, the ideologies that undergirded the knowledge-exchange process, and the supportive function of the networks.

Organizationally, both the Women's Network and the Ocooch Grazers Network constituted social movement communities, or loose associations with informal patterns that characterized how the groups functioned (Buechler 1990). The most striking organizational differences between the two networks were the geographic scale at which they were organized and the activities they pursued. The Ocooch Grazers drew its membership largely from several neighboring counties. This restricted geographic distribution permitted relatively easy travel to each other's farms for monthly pasture walks, the principal activity of the network, which focused on specific techniques related to the shared practice of rotational grazing. In contrast, participants in the Women's Network came from a much wider geographic area, meeting less often but for longer periods of time and communicating with each other between meetings through a newsletter. They had no common interest in a particular farming practice, but rather a common social location, that of woman farmer. Although unable to hold and attend meetings very frequently because of the geographic distances involved, the women still definitely benefited from the network. The differences between these two networks in the way they organized themselves suggest that farmer networks can probably function well with a variety of structures, as long as those structures are responsive to the shared needs of the participants.

Within both networks, members asserted the validity and utility of their own personal, local knowledge, even as they drew on a range of other sources of information. Such local knowledge was produced from activities and experiences that were situated in distinct physical and social locations. Emphasizing the importance of learning by doing, grass farmers steadily gained confidence in their own abilities to generate knowledge as they adapted the technique's flexible management principles to the physical characteristics of their particular farms. In the Women's Network, members drew on their own personal experiences in a gendered society and developed socially situated ways of seeing, knowing, and understanding their farming experiences. In addition to their personal knowledge of gender relations in agriculture, members of the Women's Network also articulated the importance of their own production of local knowledge for the successful practice of sustainable agricultural techniques and marketing strategies.

In these ways, the two case studies illustrate that different lived experiences seem to produce multiple and partial perspectives from which local knowledge for sustainable agriculture is generated. Rotational grazers exhibited a strong emphasis on how local knowledge was shaped by the *physical location* of the knower, as Kloppenburg (1991) predicted. By contrast, in the case of the Women's Network, it was *social location* that most profoundly influenced the unique angles of vision from which members' personal knowledge was constructed, as Feldman and Welsh (1995) anticipated.

In both networks, personal, local knowledge was extended beyond the individual to become a social product available for use and interpretation by a community of knowers. In the case of the Ocooch Grazers, grass farmers were engaged in a technological reversal of the dominant methods of dairy production, and accordingly their knowledge exchanges covered a range of technical questions, including how to develop and improve the pasture, how to produce milk in a grass-based system, and how to think about economic costs. Network events provided a valuable opportunity for members to share their own local knowledge and to learn by hearing the experiences of others, by observing how others apply the practice to their farms, and by watching others demonstrate certain techniques. Increasingly, members of the Ocooch Grazers came to believe that they themselves must produce and exchange the local knowledge that will be the foundation for grass-based dairying in the Upper Midwest.

Unlike the grass farmers who shared knowledge in order to achieve a technological reversal in conventional dairying, members of the Women's Network exchanged knowledge to accomplish a social reversal, that is, to promote the success of *women* as sustainable farmers. The knowledge-exchange process in the Women's Network encompassed members' sharing of personal experiences with encountering and overcoming gender inequity, as well as the transmittal of technical information related to agricultural production and marketing strategies. Particular emphasis was placed on gender-related issues as members shared not only their struggles but their successes, often serving as role models for one another. Feelings and emotions were productively combined with the exchange of substantive information and thus extended conventional notions of rationality that have tended to neglect those resources, as many feminists have noted. In these ways, Women's Network members conveyed a strong sense of themselves as agents of transformation, empowering and inspiring one another to move forward toward their personal goals.

Knowledge consists of both substantive information about specific topics and the ideological assumptions from which such information is created and exchanged. In both networks, members exchanged ideas, values, and beliefs about sustainable agriculture. In so doing, they demonstrated that local knowledge for sustainable agriculture can lead farmers not only to think *for* themselves as they implement alternative practices and create new enterprises but also to think *about* farming and farmers in novel ways. Consistent with the arguments put forward by Eyerman and Jamison (1991), in both networks the organizations themselves became important spaces in which new ideas and relationships developed.

Members of the Ocooch Grazers became dairy heretics as they questioned the dominant ideology of Wisconsin dairying and articulated alternative values during network interactions. Although not all members held the same views, these new ways of thinking related to rejecting the prevailing productionist ideology, creating profitable enterprises through lower costs, finding pleasure in their work, developing a sense of epistemic self-reliance, respecting their livestock, cultivating an appreciation for the environment, and empowering themselves to imagine a more successful future in agriculture. Members of the Women's Network also questioned the prevailing agricultural ideology and expressed multidimensional visions of agricultural sustainability. The primary elements of sustainability addressed by the Women's Network included the belief in the need to reduce or eliminate the use of pesticides, synthetic fertilizers, and fossil fuels; to work within the natural constraints of the landscape; to minimize input costs so as to improve farm profitability; and to build regionally based food systems that reconnect consumers and producers through new marketing arrangements.

The striking difference between the two networks with respect to ideological exchanges was the prevalence of gender discrimination issues in discussions among the members of the Women's Network. They described the ways that some of their difficulties obtaining financing, supplies, and technical assistance were caused by prejudice against their gender; that society discourages girls from obtaining skills that would enable them to farm successfully; and that women's contributions to farming are devalued or made invisible in the context of a male-dominated agricultural system. Such issues were almost never discussed at the Ocooch Grazers' gatherings even though there were women members.

Participants in both the Women's Network and the Ocooch Grazers seemed to overcome the limits of their personal knowledge by sharing, cooperating, and combining knowledge and in so doing embodied the kind of horizontal, democratic form of organizing that Wainwright (1994) has suggested is a primary characteristic of new social movements. Part of what makes these new social movements distinctive, according to Wainwright, is their critique of the hierarchical nature of what counts as valid knowledge. Such a critique was evident in the two networks studied here because they challenged the power relations in the production and distribution of agricultural knowledge by relying on their own and other members' experiential knowledge. Specifically, the grazers defied the disproportionate influence that agribusiness has over the questions that agricultural science has (and has not) asked and over the ways the answers to those questions have been expressed in technologies. Rather than continue to submit to corporate dominance of technical innovation, the grazers drew on their own collective capacities to generate and exchange the knowledge they needed. Members of the Women's Network contested male dominance of the production and distribution of social knowledge. They rejected the existing social construction that only men are capable of being farmers, and they shared knowledge about how to overcome the constraints that result from this social construction. Therefore, each network took a radical step toward building a more sustainable agriculture.

Farmers' lives and localities seem to provide a grounding for knowledge claims that are different from those grounded in the public and private institutions of agricultural research. But recognizing the contributions that farmer-generated knowledge can make to the creation of a more sustainable agriculture does not necessarily imply a rejection of other ways of knowing. In both of the networks studied here, members determined for themselves which outside resources they wanted to bring to their groups. For instance, the grazers were willing to learn what they could from the scientists and government agents who also became "believers" in the unconventional technique of rotational grazing. Similarly, members of the Women's Network were open to learning from men, albeit in an atmosphere that the women controlled and found to be more comfortable than the male-dominated settings typical of agriculture. In these ways, network members implicitly recognized that there are multiple ways of knowing the world and that different perspectives could produce different, yet partial, truths.

Thus, in forging an alternative knowledge system, neither network was trying to replace existing agricultural science with another form of knowledge production and distribution. In rejecting the all-knowing expertise characteristic of the dominant knowledge system, they created instead heterogenous knowledge systems based on horizontal exchange of information among a community of knowers with unique social and physical locations. Networks do not constitute a ready-made alternative to the dominant knowledge system, but they do suggest ways of democratizing knowledge production for sustainable agriculture. They create conditions that potentially permit mutually beneficial dialogue between these different ways of knowing so that the latter might inform each other better.

These findings indicate, however, that differential power relationships often make it hard for women to participate fully in the sustainable agriculture movement's knowledge-exchange activities (see also Hassanein 1997b). For instance, the potential heterogeneity of the Ocooch Network was limited because the women who were active members of that group did not have an equal voice in the knowledge-exchange process. Yet as the findings from the Women's Network indicate, women do have unique perspectives from which their personal knowledge is generated and exchanged. Therefore, the Ocooch network – and by extension other farmer networks – could potentially be greatly enriched by a deliberate effort to integrate a diversity of knowledge-generating perspectives in their knowledge-exchange activities. Perhaps the knowledge-exchange among the members of the Women's Network constitutes an important first step toward a more inclusive alternative knowledge base upon which a transformed, truly sustainable agriculture may be built.

Another important way in which the networks contributed toward building a sustainable agriculture was that both groups provided social support for their active members, one of the most surprising findings of this study in my view. In each case, however, members' expressed need for support arose out of the different social conditions that they encountered. As the grazers rejected the established patterns of technological development in Wisconsin dairying, they became disconnected from and even ostracized by their neighbors who continued to farm conventionally. Emphasizing the ways that adoption of rotational grazing made many into "believers" in a "religion," members of the Ocooch Network indicated that despite the practical advantages of rotational grazing there were social drawbacks to adoption of the technique. By becoming a "subculture" in the rural community, the network helped to buffer that problem as members developed a camaraderie around their common interest in the technique of rotational grazing and their common understandings of the benefits of their adoption of this unconventional practice.

Unlike the grazers, who did not talk publicly about how the network functioned for support and mentioned that aspect only in interviews with me, members of the Women's Network repeatedly stressed the importance of the supportive function. Also unlike the grazers, who needed support for doing a different kind of farming, members of the Women's Network needed support for being a different kind of farmer. In establishing a women-only space, members created what they perceived to be a less competitive and a more comfortable and respectful atmosphere than they normally encountered in the male-dominated world of agriculture. Members of the Women's

Network reasoned that such support was integral to helping women realize their personal goals in farming, which in turn would enhance the visibility of successful women in sustainable agriculture. Such proof of success, they hoped, might ultimately break down some of the social and institutional barriers that women farmers confront. If the support function seemed to be more important in the case of the Women's Network than it did to members of the Ocooch Network, perhaps it was because the grazers were trying to do an *activity* that was looked down on; but the women *themselves* were being looked down on, that is, their very gender identity was being questioned.

Despite the different social conditions that created members' need for support in each case, what becomes clear from these observations is that the support function may be crucial for sustaining – and expanding – the movement at the local level. Perhaps social movements like feminism and sustainable agriculture have underestimated the degree to which it is difficult for people to align themselves with radical social change. Yet this research shows that people clearly need support from their peers when they pursue unorthodox techniques, articulate novel ideas, and create new identities. Moreover, social support is most likely to emerge in social movement communities because these informal organizations are accessible to individuals situated in particular localities. Without the face-to-face contact possible in local and regional networks, it seems that fewer farmers will feel able to challenge established social and technological patterns in agriculture. Therefore, creating the conditions for social support at the local level appears to be an essential ingredient in advancing the goals of the sustainable agriculture movement.

Generalizing the Findings

The generalizability of these findings to sustainable farming networks elsewhere and to other social movements will be known only after further study of such networks and movements. A limited review of relevant movement literature, however, gives some reason to suspect that similar themes emerge in other sustainable farming networks. For instance, the farm improvement clubs organized by the Alternative Energy Resources Organization in Montana reportedly function as a "learning community" where sustainable agriculturalists get together to find answers to innovative questions (Matheson 1993:11). And echoing the supportive function of the farmer networks studied here, these clubs apparently reduce the isolation caused by departing from convention. As one observer explained, club members no longer "feel like fruitcakes just because they're trying to conserve resources" (quoted in Bird, Bultena, and Gardner 1995:164). Likewise, organic farmer and sustainable agriculture advocate Frederick Kirschenmann described his experiences with the Northern Plains Sustainable Agriculture Society in North Dakota. Although the original purpose of that organization was for farmers to learn about alternative farming practices from each other, participants found that "pent up in each of us was the need to share more than information. The need to share concerns, joys, fears, dreams, failures, and most of all, our common commitment to the land, was evident" (Kirschenmann 1992b:34).

In addition, women interested in sustainable agriculture in other areas of the country seem to be increasing their levels of organization for reasons that resemble those identified in my description and analysis of the Women's Network. For example, citing their needs for "inspiration" and "support," women in Minnesota formed Women in Sustainable Agriculture in late 1993. The group decided "to meet to find out new farming methods, to share their triumphs and their frustrations and to be with people who do what they do" (Johnson 1993:6; see also Chiappe 1994). Similarly, Cris Carusi (1996:2) reported that Nebraska farm women formed EQUAL, which stands for Enhanced Quality of Life, and their goal is "to learn more about sustainable farming and marketing practices, improve their leadership, communication and time management skills, and support each other while building confidence." Likewise, women involved in Practical Farmers of Iowa gathered for a first annual weekend-long conference in 1996 "to share experiences, ideas, and philosophies about sustaining land, sustaining communities, and sustaining ourselves" (Smith-Hampton 1996). In addition, Gwendolyn Ellen (1996) described how a group called Wild Women in Sustainable Agriculture was formed in Oregon in early 1996 so that women could "come together to share our knowledge, experiences, tribulations, and inspiration, and to give voice to women's perspectives of sustainable agriculture." The establishment of women-only spaces within the movement underscores criticisms voiced by Patricia Allen and Carolyn Sachs (1993) that prevailing visions of agricultural sustainability fail to incorporate or even recognize the need to change inequitable gender relations in U.S. agriculture. By gathering together in these ways, women appear interested in expanding the vision of a sustainable agriculture so that it reflects themselves and their concerns.

These examples of other sustainable farming networks give reason to speculate that processes similar to those explored in the two networks studied here may exist in other settings. Generalizations beyond the particular situations are difficult to make, however, because in-depth field research is necessarily limited to a few sites. The strength of field research methods lies not so much in the extent to which findings can be generalized than in the ability to study dynamic, small-group situations in detail, such as these farmer networks. The observational and interpretive approach employed in this study allowed me to identify what the networks meant to participants and to analyze the nature and function of their knowledge-exchange activities in the context of the whole network, a research objective that is difficult to accomplish with survey techniques. The in-depth study of particular situations also permits elaboration of existing social theory, in this case theories related to local knowledge in the sustainable agriculture movement and to knowledge exchange in social movements in general.

Perhaps more important than the methodological reasons for restricting the generalizations made from this study are the substantive reasons. That is, in formulating achievable strategies for sustainable agriculture it is necessary to take into account local and regional characteristics of agroecological conditions, labor and commodity markets, and the social organization of farming (Lighthall 1996). Gustavo Esteva (1987) argued that although there may be instructive similarities between groups that try to "regenerate people's spaces," such local efforts do not necessarily suggest a "model" that others can easily imitate or reproduce. Ultimately, sustainable agriculture must reflect the social, ecological, and economic conditions of particular localities, and, accordingly, the organizations where farmers produce and share the local knowledge that will undergird sustainable agriculture must also reflect those conditions.

Potentials and Limitations of Sustainable Farmer Networks

Recognition of the social character of local, personal knowledge in these particular facets of the sustainable agriculture movement is important in more than theoretical and empirical ways. Those committed to bringing about a sustainable agriculture will recognize that the ability to exchange local knowledge in a meaningful way with others who are similarly situated has a significant impact on the potential for social transformation. Without the capacity to pool their local knowledge and absent major transformation in the institutions of agricultural research, individual farmers experimenting with social and technical reversals in agriculture would have to invent and reinvent all techniques and ideas to achieve sustainability. While local knowledge for sustainable agriculture must be produced and applied locally, it is extended and informed considerably by direct contact with a community of knowers who share common interests. This is consistent with Wainwright's (1994:xii) emphasis on "a 'bottom up' approach to social transformation in a recognition that the knowledge shared at the base of society is essential to a socially effective and just society."

In looking to themselves and to each other for knowledge, network members are seeking creative solutions to the problems threatening the economic, ecological, and social viability of family-based farming in the Upper Midwest. The generation and exchange of local knowledge is a way of creating social change that contrasts with other approaches such as organizing for passage of legislation or trying to reorient agricultural science toward alternative lines of inquiry and research methods. "People on the farms are beginning to realize that our destiny is in our own hands," said a grass farmer who does not focus much on "who's president or secretary of agriculture or the farm bill. . . . [Instead] we try to create our own universe here. . . . It's a matter of proper networks and neighborhoods." The extent of collective action in these networks was limited to their efforts to create a space for people to interact with one another because network members tended to define the scale of social problems at a level at which they could contemplate and execute solutions on their own farms and in their own localities.

In so doing, network members did not always connect their efforts explicitly to the wider sustainable agriculture movement. For instance, while some members of the Ocooch Grazers associated their practices and ideas with sustainability, others did not. And among grazers who were self-conscious activists in the movement, there was a strategic attempt to avoid the marginal and negative connotations of sustainable agriculture in the rural community by promoting the adoption of the technique on its own merits. Indeed, the powerful economic advantages of rotational grazing seem to suggest a real possibility for expanding the number of adherents to a sustainable agricultural practice, as evidenced by the growing number of dairy farmers adopting the technique and organizing themselves in grazing networks around Wisconsin.

By comparison, most members of the Women's Network seemed to identify willingly with and explicitly support the sustainable agriculture movement. Yet a substantial number of them were reluctant to identify themselves and the organization with the feminist movement. As a result, some network leaders strategically avoided using a feminist label for the organization, fearing that it might discourage attendance of women who were suspicious or unclear about feminism. Accordingly, most members did not *explicitly* advocate for the incorporation of equal gender relations in agriculture as a goal that the sustainable agriculture movement ought to pursue more generally, even though they were clearly concerned about gender relations in agriculture and asserted their need for a women-only space. Perhaps this is not surprising because dominant visions of agricultural sustainability have not included the rectification of gender oppression as part of the picture for the future. Nevertheless, the activities of the Women's Network were specifically located in the context of the sustainable agriculture movement. Like the grazers who wanted rotational grazing to be accepted on its own merits, members of the Women's Network typically felt that a major purpose of the network was to help women demonstrate their own capacities as successful sustainable farmers and thus subvert stereotypical perceptions of women's role in agriculture through slow cultural transformations.

Whether local participants intend it or not, they seem to be a critical ingredient in advancing the goals of a wider movement by demonstrating that social change in agriculture can occur through what Karl Weick (1984) has termed "small wins" or what Wainwright (1994:80) has described as "radical gradualism." In other words, by taking incremental steps in their everyday lives, people can mobilize resources to achieve all that is presently possible in the pursuit of long-term goals. By no means did the network farmers in this study have all the answers to the problems confronting contemporary agriculture. In fact, they often seemed to discover a wide array of questions that have not been asked before and will not be answered easily. Nonetheless, in developing and exchanging the knowledge that might potentially undergird the creation of a more sustainable agriculture in the Upper Midwest, network members acted on Harriet Friedmann's (1993:228) assertion that the "promising solution lies in locality and seasonality." Furthermore, they offered each other much-needed support as they pursued activities and identities that were marginalized by others in the agricultural community.

The farmer networks in this study did not act in social isolation as the members pursued their personal and collective goals. Academics, corporations, and government agencies have all established connections to the sustainable farming networks. The influence of a wider culture that presumably holds different values can potentially undermine the autonomy of the networks and the direction their knowledge development takes. To date that autonomy appears to be respected, and as Wainwright (1994:278) pointed out, "Autonomy is the basis of a relationship, not the synonym of separateness." I agree and suggest that those activists and academics who support the goals of the sustainable agriculture movement actively seek ways to enhance the strength of the farmer networks without undermining the networks' autonomy. This can be done, for example, by affirming the importance of local and site-specific research and inquiry and by continuing and expanding efforts to connect people across spheres of influence. For network members, retaining control over the development and dissemination of local knowledge will require a deliberate effort to maintain and defend the autonomy of the spaces in which they exchange knowledge and build community.

The power and promise of sustainable farming networks lies with the observation that problem solving through collective creation and exchange of knowledge is the foundation of our democratic society. As Carl C. Taylor (1941:7) concluded so many years ago: "We believe in democracy because we believe that every individual has a contribution to make to the solution of our common problems. A meeting of neighbors and friends . . . is the grass roots of democratic organization, and a trading of ideas among neighbors is the way to make democracy work."

Sustainable Farmer Networks in Wisconsin, 1986–95

1. Apple River Sustainable Network Advisory Group
2. Central Sands Grazing Network
3. Central Wisconsin River Graziers
4. Chippewa Valley Profitability Network
5. Coulee Graziers
6. Dane-Green Graziers
7. Eastern Wisconsin Sustainable Farmers Network
8. Fond-O-Grass Graziers
9. Grant County Graziers
10. Great Lakes Basin Intensive Rotational Grazing Network
11. Great River Graziers
12. Integrated Pest Management and Pest Reporting Program for Wisconsin Ornamental Nurseries
13. Iowa County Graziers Network
14. Kickapoo Organic Resource Network
15. Lafayette County Rotational Graziers Network
16. Madison Area Community Supported Agriculture Coalition
17. Marsh Graziers
18. Minnesota–Western Wisconsin Community Farm Association
19. North Central Graziers Network
20. Northeastern Wisconsin Sustainable Farmers Network
21. Northland Graziers
22. Northwest Wisconsin Graziers Network
23. Ocooch Grazers
24. Sauk County Grazing Network
25. Southwest Wisconsin Farmers Research Network
26. Sustainable Farmers Education Network

27. Waupaca County Grazing Group
28. Western Wisconsin Sustainable Farming Network
29. Wisconsin Small-Scale Chicken Farmers Network
30. Wisconsin Women's Sustainable Farming Network

Interview Guide for Ocooch Grazers Network

I thought we could start out by talking a bit about your thoughts on farming and your farming background. How long have you been farming on this farm?

Before you came here, did you have any previous farming experience?
If yes: What kind of experience did you have?

Why do you farm rather than make a living another way?
Probe: What's the main reason you farm?

How would you briefly describe this farm and farming operation that you are part of now? In other words, how would you characterize it? (E.g., how many cows do you milk, are you doing seasonal milking? If not now, are you considering it? Have you built or are you considering a low-cost parlor?)

When did you first begin to practice rotational grazing?

Why do you practice rotational grazing?
Probe: Any other good things about it that you can think of?

Are there drawbacks to practicing rotational grazing?
Probe: Any other disadvantages to it that you can think of?

Do you see grazers as part of a larger movement for change in agriculture?
Probe: I'm wondering if the increased interest in rotational grazing is a technical change in how we do dairying, or if it implies bigger changes with broader implications. Do you have any ideas on that?

How did you first learn about the possibility of using rotational grazing in a dairy operation? After you decided to adopt it, how did you figure out how to set up the pasturing system on your farm?

Think back over the whole time you have been grazing – from the very beginning up until now. What things have you relied on most in learning how to do rotational grazing on your farm?

Probe: Note which of these were mentioned, then ask about those not mentioned and whether they have been useful or played a role in learning:
· *books, magazines, newspaper articles*
· *trial and error; experience, observation*
· *other individual farmers practicing grazing*
· *going to network events or pasture walks*
· *extension agents*
· *other government agents*
· *other people associated with the university*
· *agribusiness*
· *consultants*

I hear a lot of people say that there are no "recipes" to grazing or that there is no "prescribed formula" on how to graze. Would you explain to me what is meant by that? Has that been true for you?

Now I would like to focus a little on the Ocooch Grazers Network. How did you first become involved in the network?

What do you personally get out of participating in the network? (make a note of the things they mention).

Probes: Think about the last meeting you were at. What would you say you got out of it? Any other things you can think of that you get out of belonging to the network?

Ask women only: Do you see any differences in the way women participate in the network and the way men participate or in the usefulness of the network to women compared with men? Do you have any ideas on that?

It seems that on many topics related to grass-based dairying there are often many different ideas, sometimes that are contradictory, being passed around among farmers about what works or doesn't work. When there is contradictory information like that, how do you decide what to do?

How do you think these kinds of issues – where people have different ideas about what should be done – will be resolved for the grazing community as a whole?

Probe: I'm wondering if these different ideas will just remain or if in the long run there will be agreement on them somehow.

One of the other things I am interested in is what your views are on the role that agricultural research and extension play in Wisconsin dairying. How would you evaluate their role?
Probe: Is it positive and in what ways? Is it negative and in what ways?

Rotational grazing is often referred to as a technique associated with sustainable agriculture. Do you think that is accurate? Why? or why not?

How would you define sustainable agriculture?

What would you say the purposes of the network are?

What direction would you like to see the network go in the future?
Probe: Would you like to see the network continue the way it has been or are there things that you would like to see the network do differently in the future?

Is there anything else you would like to bring up that we haven't discussed or questions that you have for me?

Interview Guide for Women's Network

I thought we could start out by talking a bit about your thoughts on farming and your farming background. How long have you been farming on this farm?

Before you came here, did you have any previous farming experience?
If yes: What kind of experience did you have?

Why did you start farming? Have your reasons for farming changed since you began?
Probe: In other words, what are the main reasons you farm (rather than make a living another way)?

How would you describe this farm and the farming-related enterprises that you are part of now?

How do you market your products?

Would you describe yourself as the primary farmer of this farm?
Probe: Who does what work on the farm? Do you supply all the labor on the farm? Do you share decision making with a partner?

What are your goals for your farm?
Probe: Any other goals you can think of?

When we talked about the mission statement of the Women's Network, one of the things that we discussed is sustainable farming. Let's talk about that for a few minutes. First, how would you define sustainable agriculture?

Are there things you do now on this farm that you consider to be associated with sustainable agriculture? (E.g., farm practices, ways you handle marketing your products)

If yes, probe: Do you do these things because they are sustainable or are there other reasons?

Are there things you associate with sustainable agriculture that you would like to do in the future? Why aren't you doing that or those things yet?

Let's talk a little about some of the things you mentioned. [Ask the following about one or more of the practices, marketing strategies, or other aspects of the operation people identified above (i.e., things they are doing currently).]

a) How did you originally learn about ————?
Probe: Do you remember how you got that idea?

b) What things have you relied on most in learning how to ————?
Probes: Note which of these were mentioned, then ask about those not mentioned and whether they have been useful or played a role in learning:
· books, magazines, newspaper articles
· trial and error; experience, observation
· other individual farmers
· going to network events or pasture walks
· conferences
· extension agents
· other government agents
· other people associated with the university
· agribusiness

One of the other things I am interested in is what your views are on the role that agricultural research and extension play in Wisconsin agriculture. How would you evaluate their role?
Probes: Is it positive and in what ways? Is it negative and in what ways?

Now let's talk a little about the Women's Network. How did you first become involved in the network?

What do you personally get out of participating in the Women's Network?
Probes: What would you say you got out of the last meeting you were at? Any other things you can think of that you get out of belonging to the network?

How would you compare the things you learn in the Women's Network to other information sources available to farmers?

One of the things that a lot of women in the network stress is the value of the support they find there. Is getting support from the network important to you? If so, in what ways?

A related idea that some have expressed is the feeling of being inspired by the other women in the network. What do you think is meant by that? Have you been inspired by the women there? If so, in what way?

I'd like us to talk some about your views and experiences of being a woman farmer. Are there advantages to being a woman farmer? Are there disadvantages or obstacles you encounter as a woman farmer?

How do you see this changing or being overcome? Should it be changed?
Do you think people promoting sustainable agriculture can have a role in addressing these issues?

What do you think might be the consequences of addressing some of the issues women farmers confront?

What direction would you like to see the network go in the future? Are there things you would like to see the network do that it is not currently doing?

Are there any other things you would like to bring up or any questions you want to ask me?

References

Alinsky, Saul D. 1971. *Rules for radicals: A pragmatic primer for realistic radicals*. New York: Vintage Books.

Allen, Patricia, and Carolyn Sachs. 1993. Sustainable agriculture in the United States: Engagements, silences, and possibilities for transformation. Chap. 6 in *Food for the future: Conditions and contradictions of sustainability*. Edited by Patricia Allen. New York: Wiley.

Allen, Patricia, Debra Van Dusen, Jackelyn Lundy, and Stephen Gliessman. 1991. Integrating social, environmental, and economic issues in sustainable agriculture. *American Journal of Alternative Agriculture* 6 (1): 34–39.

Anderson, Kathryn, and Dana C. Jack. 1991. Learning to listen: Interview techniques and analyses. Chap. 1 in *Women's words: The feminist practice of oral history*. Edited by Sherna Berger Gluck and Daphne Patai. New York: Routledge.

Appropriate Technology Transfer for Rural Areas (ATTRA). 1998. ATTRA celebrates tenth anniversary. *ATTRA News* 6 (3): 5.

Barlett, Peggy F. 1993. *American dreams, rural realities: Family farms in crisis*. Chapel Hill: University of North Carolina Press.

Beeman, Randal. 1993. The trash farmer: Edward Faulkner and the origins of sustainable agriculture in the United States, 1943–1953. *Journal of Sustainable Agriculture* 4 (1): 91–102.

Benford, Robert D. 1992. Social movements. Vol. 4 of *Encyclopedia of Sociology*. Edited by Edgar F. Borgatta and Marie L. Borgatta. New York: Macmillan.

Bennett, John W. 1986. Research on farmer behavior and social organization. Chap. 16 in *New directions in agriculture and agricultural research: Neglected dimensions and emerging alternatives*. Edited by Kenneth A. Dahlberg. Totowa NJ: Rowman & Allenheld.

Bernard, Jessie. 1981. *The female world*. New York: Free Press.

Berry, Wendell. 1977. *The unsettling of America: Culture and agriculture*. New York: Avon Books.

———. 1984. Whose head is the farmer using? Whose head is using the farmer? Chap. 2 in *Meeting the expectations of the land: Essays in sustainable agriculture and stewardship*. Edited by Wes Jackson, Wendell Berry, and Bruce Colman. San Francisco: North Point Press.

Beus, Curtis E., and Riley E. Dunlap. 1990. Conventional versus alternative agriculture: The paradigmatic roots of the debate. *Rural Sociology* 55 (4): 590–616.

Bezdicek, David F., and Colette DePhelps. 1994. Innovative approaches for integrated research and educational programs. *American Journal of Alternative Agriculture* 9 (1, 2): 3–8.

Bird, Elizabeth Ann, Gordon L. Bultena, and John C. Gardner, eds. 1995. *Planting the future: Developing an agriculture that sustains land and community*. Ames: Iowa State University Press.

Bogdan, Robert, and Steven J. Taylor. 1975. *Introduction to qualitative research methods: A phenomenological approach to the social sciences*. New York: Wiley.

Bonanno, Alessandro, Lawrence Busch, William Friedland, Lourdes Gouveia, and Enzo Mingione, eds. 1994. *From Columbus to ConAgra: The globalization of agriculture and food*. Lawrence: University Press of Kansas.

Borland, Katherine. 1991. "That's not what I said": Interpretive conflict in oral narrative research. Chap. 4 in *Women's words: The feminist practice of oral history*. Edited by Sherna Berger Gluck and Daphne Patai. New York: Routledge.

Boyte, Harry C. 1984. *Community is possible: Repairing America's roots*. New York: Harper & Row.

Braverman, Harry. 1974. *Labor and monopoly capital: The degradation of work in the twentieth century*. New York: Monthly Review Press.

Briggs, Charles. 1986. *Learning how to ask: A sociolinguistic appraisal of the role of the interview in social science research*. Cambridge: Cambridge University Press.

Brokensha, David. 1989. Local management systems and sustainability. In *Food and farm: Current debates and policies*. Monographs in Economic Anthropology 7. Edited by Christina Gladwin and Kathleen Truman. Lanham MD: University Press of America.

Buechler, Steven M. 1990. *Women's movements in the United States: Woman suffrage, equal rights, and beyond*. New Brunswick: Rutgers University Press.

Busch, Lawrence, and William B. Lacy. 1983. *Science, agriculture, and the politics of research*. Boulder: Westview Press.

Buttel, Frederick H. 1993a. Ideology and agricultural technology in the late twentieth century: Biotechnology as symbol and substance. *Agriculture and Human Values* (spring): 5–15.

————. 1993b. The production of agricultural sustainability: Observations from the sociology of science and technology. Chap. 1 in *Food for the future: Conditions and contradictions of sustainability*. Edited by Patricia Allen. New York: Wiley.

Buttel, Frederick H., and Gilbert W. Gillespie Jr. 1988. Agricultural research and development and the appropriation of progressive symbols: Some observations on the politics of ecological agriculture. *Cornell Rural Sociology Bulletin Series* 151.

Buttel, Frederick H., Olaf F. Larson, and Gilbert W. Gillespie Jr. 1990. *The sociology of agriculture*. Westport CT: Greenwood Press.

Campbell, David. 1993. The economic and social viability of rural communities: BGH vs. rotational grazing. Chap. 7 in *The dairy debate: Consequences of bovine growth hormone and rotational grazing technologies*. Edited by William C. Liebhardt. Davis: University of California, Sustainable Agriculture Research and Education Program.

Caneff, Denny. 1993. *Sustaining land, people, animals, and communities: The case for livestock in a sustainable agriculture*. Washington DC: Midwest Sustainable Agriculture Working Group.

Carson, Rachel. 1962. *Silent spring*. New York: Fawcett Crest.

Carusi, Cris. 1996. Women's group enhances members' quality of life. *Nebraska Sustainable Agriculture Society Newsletter* 55:2.

Chambers, Robert. 1983. *Rural development: Putting the last first*. New York: Wiley.

Chiappe, Marta Beatriz. 1994. Women in sustainable agriculture: A study of Minnesota family farms. Ph.D. dissertation, University of Minnesota.

Christie, Margaret M. 1996. Carl C. Taylor, "organic intellectual" in the New Deal Department of Agriculture. Master's thesis, University of Wisconsin–Madison.

Clancy, Katherine L. 1993. Sustainable agriculture and domestic hunger: Rethinking a link between production and consumption. Chap. 11 in *Food for the future: Conditions and contradictions of sustainability*. Edited by Patricia Allen. New York: Wiley.

Cobb, John B. 1984. Theology, perception, and agriculture. Chap. 13 in *Agricultural sustainability in a changing world order*. Edited by Gordon K. Douglass. Boulder: Westview Press.

Cochrane, Willard W. 1979. *The development of American agriculture: A historical analysis*. Minneapolis: University of Minnesota Press.

Cohen, Jean L. 1985. Strategy or identity: New theoretical paradigms and contemporary social movments. *Social Research* 52 (4): 663–716.

Crouch, Martha L. 1995. Why science can't save the Earth. Unpublished manuscript, Indiana University.

Danbom, David B. 1986. Publicly sponsored agricultural research in the United States from an historical perspective. Chap. 6 in *New directions in agriculture and agricultural research: Neglected dimensions and emerging alternatives*. Edited by Kenneth A. Dahlberg. Totowa NJ: Rowman & Allenheld.

DeLind, Laura B. 1994. Organic farming and social context: A challenge for us all. *American Journal of Alternative Agriculture* 9 (4): 146–47.

Diani, Mario. 1992. The concept of social movement. *Sociological Review* 40 (1): 1–25.

Downey, Gary L. 1986. Ideology and the Clamshell identity: Organizational dilemmas in the anti-nuclear power movement. *Social Problems* 33 (5): 357–73.

Ehlers, Tracy Bachrach. 1987. The matrifocal farm. Chap. 5 in *Farm work and fieldwork: American agriculture in anthropological perspective*. Edited by Michael Chibnik. Ithaca: Cornell University Press.

Ellen, Gwendolyn. 1996. What Wild Women in Sustainable Agriculture (WWISA) do: A spirited invitation and update. *In Good Tilth* 7 (7): 4.

Enshayan, Kamyar, Deb Stinner, and Ben Stinner. 1992. Farmer to farmer. *Journal of Soil and Water Conservation* 47 (2): 127–30.

Esbjornson, Carl D. 1992. Once and future farming: Some meditations on the historical and cultural roots of sustainable agriculture in the United States. *Agriculture and Human Values* 9 (3): 20–30.

Esteva, Gustavo. 1987. Regenerating people's spaces. *Alternatives* 12:125–52.

Evans, Sara M., and Harry C. Boyte. 1986. *Free spaces: The sources of democratic change in America*. New York: Harper & Row.

Eyerman, Ron, and Andrew Jamison. 1991. *Social movements: A cognitive approach*. University Park: Pennsylvania State University Press.

Fales, Steven L. 1994. Rethinking the role of pasture for dairy cows. *American Forage and Grassland Council News* 5 (3): 3–6.

Fals Borda, Orlando. 1988. *Knowledge and social movements*. Regents Lectures, Latin American Studies Program, University of California, Santa Cruz. Santa Cruz: Merrill Publications.

Fee, Elizabeth. 1986. Critiques of modern science: The relationship of feminism to other radical epistemologies. Chap. 3 in *Feminist approaches to science*. Edited by Ruth Blier. New York: Pergamon Press.

Feldman, Shelley, and Rick Welsh. 1995. Feminist knowledge claims, local knowledge, and gender divisions of agricultural labor: Constructing a successor science. *Rural Sociology* 60 (1): 23–43.

Ferree, Myra Marx, and Beth B. Hess. 1985. *Controversy and coalition: The new feminist movement*. Boston: Twayne.

Fine, Michelle, and Virginia Vanderslice. 1992. Qualitative activist research: Reflections on methods and politics. Chap. 10 in *Methodological issues in applied social psychology*. Edited by Fred Boyd Bryant, John Edwards, R. Scott Tindale, Emil J. Posavac, Linda Heath, Eaaron Henderson, and Yoland Suarez-Balcazar. New York: Plenum Press.

Flora, Cornelia Butler. 1992. Reconstructing agriculture: The case for local knowledge. *Rural Sociology* 57 (1): 92–97.

Francis, Charles A. 1990. Sustainable agriculture: Myths and realities. *Journal of Sustainable Agriculture* 1 (1): 97–106.

Freire, Paulo. 1970. *Pedagogy of the oppressed*. New York: Continuum.

Freudenberger, C. Dean. 1986. Value and ethical dimensions of alternative agricultural approaches: In quest of a regenerative and just agriculture. Chap. 15 in *New directions for agriculture and agricultural research: Neglected dimensions and emerging alternatives*. Edited by Kenneth A. Dahlberg. Totowa NJ: Rowman & Allanheld.

Frey, Lawrence R. 1994. The naturalistic paradigm: Studying small groups in the postmodern era. *Small Group Research* 25 (4): 551–77.

Friedland, William H. 1978. *Social sleepwalkers*. Research Monograph 13. Davis: Department of Applied and Behavioral Sciences, University of California–Davis.

Friedland, William H., and Tim Kappel. 1979. *Production or perish: Changing the inequities of agricultural research priorities*. Santa Cruz: Project of Social Impact Assessment and Values, University of California–Santa Cruz.

Friedmann, Harriet. 1993. After Midas's feast: Alternative food regimes for the future. Chap. 9 in *Food for the future: Conditions and contradictions for sustainability*. Edited by Patricia Allen. New York: Wiley.

Fyksen, Jane. 1994a. New sustainable farm network for "women only." *Agri-view* 20 (7, section 2): 2.

———. 1994b. Sustainable ag program has emotional support, but finances running out. *Agri-view* 20 (37, section 2): 2.

Gage, Susan, and Stewart Smith. 1989. The Moore dairy farm. *American Journal of Alternative Agriculture* 4 (1): 35–37.

Gale-Sinex, Michele. 1994. CIAS forms grazing-based dairy systems interest group; project teams moving ahead with four work clusters. *CIAS Connections*. (Newsletter of the Center for Integrated Agricultural Systems, University of Wisconsin –Madison) 4 (1): 1–2, 10.

Galloway, Jennifer. 1995. Family farm means big business. *Wisconsin State Journal* (January 29): 1E–2E.

Garkovich, Lorraine, and Janet Bokemeier. 1988. Agricultural mechanization and American farm women's economic roles. Chap. 12 in *Women and farming: Changing roles, changing structures*. Edited by Wava G. Haney and Jane B. Knowles. Boulder: Westview Press.

Geertz, Clifford. 1973. Thick description: Toward an interpretive theory of culture. Chap. 1 in *The interpretation of cultures: Selected essays*. New York: Basic Books.

Gerber, John M. 1992. Farmer participation in research: A model for adaptive research and education. *American Journal of Alternative Agriculture* 7 (3): 118–21.

Gottlieb, Robert. 1993. *Forcing the spring: The transformation of the American environmental movement*. Washington DC: Island Press.

Haney, Wava G., and Jane B. Knowles. 1988. Making "the invisible farmer" visible. Introduction to *Women and farming: Changing roles, changing structures*. Edited by Wava G. Haney and Jane B. Knowles. Boulder: Westview Press.

Haraway, Donna. 1988. Situated knowledges: The science question in feminism and the privilege of partial perspective. *Feminist Studies* 14 (3): 575–99.

Harding, Sandra. 1986. *The science question in feminism*. Ithaca: Cornell University Press.

Harper, Douglas. 1987. *Working knowledge: Skill and community in a small shop*. Berkeley: University of California Press.

Harwood, Richard. 1993. A look back at USDA's Report and Recommendations on Organic Farming. *American Journal of Alternative Agriculture* 8 (4): 150–54.

Hassanein, Neva. 1997a. Exchanging knowledge, building community: Farmer networks and the sustainable agriculture movement. Ph.D. dissertation, University of Wisconsin–Madison.

———. 1997b. Networking knowledge in the sustainable agriculture movement: Some implications of the gender dimension. *Society and Natural Resources* 10 (3): 251–57.

Hassanein, Neva, and Jack R. Kloppenburg Jr. 1995. Where the grass grows again: Knowledge exchange in the sustainable agriculture movement. *Rural Sociology* 60 (4): 721–40.

Hemken, Douglas. 1995. When farmers were scientists: Varieties of social demarcation and organization in 19th century agricultural science. Unpublished manuscript, University of Wisconsin–Madison.

Henry A. Wallace Institute for Alternative Agriculture. 1992. The Wallace Institute's agenda for the future. *Annual Report*. Greenbelt MD: Henry A. Wallace Institute for Alternative Agriculture.

Hewitt, Tracy Irwin, and Katherine R. Smith. 1995. *Intensive agriculture and environmental quality: Examining the newest agricultural myth*. Greenbelt MD: Henry A. Wallace Institute for Alternative Agriculture.

Hightower, Jim. 1976. Hard tomatoes, hard times: The failure of the land grant college complex. Chap. 7 in *Radical agriculture*. Edited by Richard Merrill. New York: New York University Press.

Howard, Sir Albert. 1945. Introduction to *Pay dirt: Farming and gardening with composts,* by J. I. Rodale. New York: Devin-Adair.

Jackson, Wes. 1987. *Altars of unhewn stone: Science and the Earth.* New York: North Point Press.

———. 1990. Agriculture with nature as analogy. Chap. 14 in *Sustainable agriculture in temperate zones.* Edited by Charles A. Francis, Cornelia Butler Flora, and Larry D. King. New York: Wiley.

Jackson-Smith, Douglas, Bradford Barham, Monica Nevius, and Rick Klemme. 1996. *Grazing in dairyland: The use and performance of management intensive rotational grazing among Wisconsin dairy farms.* Technical Report 5. Madison: Agricultural Technology and Family Farm Institute.

Jellison, Katherine. 1993. *Entitled to power: Farm women and technology, 1913– 1963.* Chapel Hill: University of North Carolina Press.

Jensen, Joan M. 1981. *With these hands: Women working on the land.* Old Westbury NY: Feminist Press.

Johnson, Dawn. 1993. This land is their land: Women in sustainable agriculture find inspiration and support in talking freely about their way of life. *Northfield News,* 14 October, Harvest Edition, pp. 6–7.

Keller, Evelyn Fox. 1987. The gender/science system: Or is sex to gender as nature is to science? *Hypatia* 2 (3): 37–49.

Kirkendall, Richard S. 1987. Up to now: A history of American agriculture from Jefferson to revolution to crisis. *Agriculture and Human Values* 4 (1): 4–26.

Kirschenmann, Frederick. 1992a. Beyond the second phase of sustainable agriculture. Paper presented at the 1992 Organic Farming Symposium at Asilomar CA.

———. 1992b. What can alternative farm systems and rural communities do for each other? In *Alternative farming systems and rural communities: Exploring the connections.* Symposium Proceedings. Chevy Chase MD: Henry A. Wallace Institute.

———. N.d. Our purpose is to serve: Challenges of the 21st century. Unpublished manuscript.

Kloppenburg, Jack R. Jr. 1988. *First the seed: The political economy of plant biotechnology, 1492–2000.* New York: Cambridge University Press.

———. 1991. Social theory and the de/reconstruction of agricultural science: Local knowledge for an alternative agriculture. *Rural Sociology* 56 (4): 519–48.

———. 1992. Science in agriculture: A reply to Molnar, Duffy, Cummins, and Van Santen and to Flora. *Rural Sociology* 57 (1): 98–107.

Knorr-Cetina, Karin. 1984. The fabrication of facts: Toward a microsociology of scientific knowledge. Chap. 9 in *Society and knowledge.* Edited by Nico Stehr and Volker Mega. New Brunswick NJ: Transaction Books.

Knowles, Jane. 1985. Science and farm women's work: The agrarian origins of home economic extension. *Agriculture and Human Values* 2 (1): 52–55.

Krcil, Larry, and Shawn Gralla. 1995. *Udder sense: Low-cost, sustainable strategies of resourceful dairy farmers*. Walthill NE: Center for Rural Affairs.

Krome, Margaret. 1988. *The Southwest Wisconsin Farmers' Research Network, 1986–1987: A two-year case history of an on-farm research project*. Mt. Horeb: Wisconsin Rural Development Center.

Lacy, William B. 1993. Can agricultural colleges meet the needs of sustainable agriculture? *American Journal of Alternative Agriculture* 8 (1): 40–45.

Lareau, Annette. 1989. *Home advantage: Social class and parental intervention in elementary education*. London: Palmer Press.

Larsen, H. J., and R. F. Johannes. 1965. *Summer forage: Stored feeding, green feeding, and strip grazing*. Wisconsin Agricultural Experiment Station Research Bulletin 257. Madison: University of Wisconsin.

Lasley, Paul, Eric Hoiberg, and Gordon Bultena. 1993. Is sustainble agriculture an elixir for rural communities? *American Journal of Alternative Agriculture* 8 (3): 133–39.

Lather, Patti. 1991. *Getting smart: Feminist research and pedagogy with/in the postmodern*. New York: Routledge.

Latour, Bruno, and Steve Woolgar. 1986. *Laboratory life: The construction of scientific facts*. Princeton: Princeton University Press.

Lawrence, Kathy. 1995. A campaign for sustainability. *In Context* 42: 48–49.

Leiss, William. 1972. *The domination of nature*. Boston: Beacon Press.

Leopold, Aldo. [1949] 1970. *A Sand County almanac*. New York: Ballantine Books.

Lévi-Strauss, Claude. 1962. *The savage mind*. Chicago: University of Chicago Press.

Liebhardt, William C. 1993. Farmer experience with rotational grazing: A case study approach. Chap. 4 in *The dairy debate: Consequences of bovine growth hormone and rotational grazing technologies*. Edited by William C. Liebhardt. Davis: University of California, Sustainable Agriculture Research and Education Program.

Lighthall, David R. 1996. Sustainable agriculture in the Corn Belt: Production-side progress and demand-side constraints. *American Journal of Alternative Agriculture* 11 (4): 168–74.

Lockeretz, William, and Molly D. Anderson. 1993. *Agricultural research alternatives*. Lincoln: University of Nebraska Press.

Logsdon, Gene. 1984. The importance of traditional farming practices for sustainable modern agriculture. Chap. 1 in *Meeting the expectations of the land: Essays in sustainable agriculture and stewardship*. Edited by Wes Jackson, Wendell Berry, and Bruce Colman. San Francisco: North Point Press.

Marcus, Alan I. 1985. *Agricultural science and the quest for legitimacy*. Ames: Iowa State University Press.

Matheson, Nancy. 1993. AERO farm improvement clubs: A collaborative learning community. *Journal of Pesticide Reform* 13 (1): 11–13.

―――. 1997. AERO fosters change in agricultural institutions. *AERO Sun Times: The Newsletter of the Alternative Energy Resources Organization* 24 (1): 11, 15.

McAdam, Doug, John D. McCarthy, and Mayer N. Zald. 1996. Opportunities, mobilizing structures, and framing processes – toward a synthetic, comparative perspective on social movements. Introduction to *Comparative perspectives on social movements: Political opportunities, mobilizing structures, and cultural framings*. Edited by Doug McAdam, John D. McCarthy, and Mayer N. Zald. New York: Cambridge University Press.

McCorkle, Constance M. 1989. Toward a knowledge of local knowledge and its importance for agricultural RD&E. *Agriculture and Human Values* 6 (3): 4–12.

McNair, Joel. 1992a. The "new" grazing: Where dairy heresy is spoken. *Agri-view* 18 (March 5): 1.

―――. 1992b. Grandpa never had it this good. *Agri-view: Special Edition on the New Dairy Grazing* (February): 1.

―――. 1992c. On pasture, UW isn't leading the way. *Agri-view: Special Edition on the New Dairy Grazing* (February): 14.

―――. 1993. Updating our grass views. *Agri-view* 19 (18, section 2): 1–2.

―――. 1994. On-farm research yielding results. *Grazing* (Special publication of *Agri-view*) 1 (3): 3.

―――. 1995. State budget drops Sustainable Ag Program. *Agri-view* 21 (34): 5.

Meine, Curt. 1987. The farmer as conservationist: Aldo Leopold on agriculture. *Journal of Soil and Water Conservation* 42 (3): 144–49.

Melucci, Alberto. 1985. The symbolic challenge of contemporary movements. *Social Research* 52 (4): 789–816.

Merchant, Carolyn. 1980. *The death of nature: Women, ecology and the scientific revolution*. New York: Harper & Row.

―――. 1992. *Radical ecology: The search for a livable world*. New York: Routledge.

Mooney, Patrick H., and Theo J. Majka. 1995. *Farmers' and farm workers' movements: Social protest in American agriculture*. New York: Twayne.

Moses, Marion. 1993. Farmworkers and pesticides. Chap. 10 in *Confronting environmental racism: Voices from the grassroots*. Edited by Robert D. Bullard. Boston: South End Press.

Mulkay, Michael. 1979. *Science and the sociology of knowledge*. London: George Allan & Unwin.

Murphy, Bill M. 1991. *Greener pastures on your side of the fence: Better farming with Voisin grazing management*. 2d ed. Colchester VT: Arriba.

Murphy, Bill M., and John R. Kunkel. 1993. Sustainable agriculture: Controlled grazing vs. confinement feeding of dairy cows. Chap. 3 in *The dairy debate: Consequences of bovine growth hormone and rotational grazing technologies.* Edited by William C. Liebhardt. Davis: University of California, Sustainable Agriculture Research and Education Program.

National Research Council (NRC). 1989. *Alternative agriculture.* Washington DC: National Academy Press.

Neth, Mary. 1995. *Preserving the family farm: Women, community, and the foundations of agribusiness in the Midwest, 1900–1940.* Baltimore: Johns Hopkins University Press.

Orr, David. 1996. Slow knowledge. *Conservation Biology* 10 (3): 699–702.

Ostrom, Marcia. 1997. Toward a community supported agriculture: A case study of resistance and change in the modern food system. Ph.D. dissertation, University of Wisconsin–Madison.

Patton, Michael Quinn. 1990. *Qualitative evaluation and research methods.* Newbury Park CA: Sage.

Peters, Suzanne. 1979. Organic farmers celebrate organic research: A sociology of popular science. Chap. 12 in *Counter-movements in the sciences.* Edited by Helga Nowotny and Hilary Rose. Dordrecht, Holland: D. Reidel.

Petulla, Joseph M. 1977. *American environmental history: The exploitation and conservation of natural resources.* San Francisco: Boyd & Fraser.

Polanyi, Michael. 1966. *The tacit dimension.* Gloucester MA: Peter Smith.

Prus, Robert. 1992. Producing social science: Knowledge as a social problem in academia. *Perspectives on Social Problems* 3:57–78.

Pulvermacher, Carl. 1993. Grazing tools from down under: Labor-saving strategies turn forage into low-cost milk. *New Farm* 15 (5): 24–26.

Reinharz, Shulamit. 1984. *On becoming a social scientist: From survey research and participant observation to experiential analysis.* New Brunswick NJ: Transaction Books.

Richards, Paul. 1993. Cultivation: Knowledge or performance? Chap. 3 in *Anthropology of development.* Edited by Mark Hobart. London: Routledge.

Rittmann, Stephanie. 1994. Exploring minds on changing farms: A case study of a grass farmer network in southwestern Wisconsin. Master's thesis, University of Wisconsin–Madison.

Rose, Hilary. 1983. Hand, brain, and heart: A feminist epistemology for the natural sciences. *Signs: Journal of Women in Culture and Society* 9 (11): 73–90.

Rosenfeld, Rachel Ann. 1985. *Farm women: Work, farm, and family in the United States.* Chapel Hill: University of North Carolina Press.

Rosmann, Ronald L. 1994. Farmer initiated on-farm research. *American Journal of Alternative Agriculture* 9 (1, 2): 34–37.

Rowe, J. Stan. 1990. *Home place: Essays on ecology*. Edmonton, Alberta: NeWest Publishers.

Rust, J. W., C. C. Shaeffer, V. R. Eidman, R. D. Moon, and R. D. Mathison. 1995. Intensive rotational grazing for dairy cattle feeding. *American Journal of Alternative Agriculture* 10 (4): 147–51.

Sachs, Carolyn E. 1983. *The invisible farmers: Women in agricultural production*. Totowa NJ: Rowman & Allanheld.

———. 1996. *Gendered fields: Rural women, agriculture, and environment*. Boulder: Westview Press.

Savory, Allan. 1988. *Holistic resource management*. Covelo CA: Island Press.

Schaefer, Paul. 1993. WWSFN closes out first year: Coordinator's comments. *News for the Western Wisconsin Sustainable Farming Network* 2 (1): 1.

Schor, Joel. 1992. Fantasy and reality: The black farmer's place in American agriculture. *Agriculture and Human Values* 9 (1): 72–78.

———. 1996. Black farmers/farms: The search for equity. *Agriculture and Human Values* 13 (3): 48–63.

Scott, Alan. 1990. *Ideology and the new social movements*. London: Unwin Hyman.

Shirley, Christopher. 1993. Milking for money or for profit? *New Farm Magazine* September–October: 31–34.

Shiva, Vandana. 1989. *Staying slive: Women, ecology and development*. London: Zed Books.

Singleton, Royce A. Jr., Bruce C. Straits, and Margaret Miller Sraits. 1993. *Approaches to social research*. 2d ed. New York: Oxford University Press.

Smith, Dorothy E. 1987. *The everyday world as problematic: A feminist sociology*. Boston: Northeastern University Press.

Smith-Hampton, Margaret. 1996. PFI women's 1996 winter gathering. *Practical Farmer: Quarterly Newsletter of Practical Farmers of Iowa* 11 (1): 20.

Stevenson, Steve, and Rick Klemme. 1991. CIAS models public input for land grant universities. *CIAS Connections* (Newsletter of Center for Integrated Agricultural Systems, University of Wisconsin–Madison) 2 (1): 1–3, 8.

Stoecker, Randy. 1995. Community, movement, organization: The problem of identity convergence in collective action. *Sociological Quarterly* 36 (1): 111–30.

Strange, Marty. 1988. *Family farming: A new economic vision*. Lincoln: University of Nebraska Press.

Strauss, Anselm, and Juliet Corbin. 1990. *Basics of qualitative research: Grounded theory procedure and techniques*. Newbury Park CA: Sage.

Suppe, Frederick. 1987. The limited applicability of agricultural research. *Agriculture and Human Values* 5 (4): 4–14.

Sustainable Agriculture Working Group (SAWG). N.d. *Midwest Sustainable Agriculture Working Group*. Brochure.

Taylor, Carl C. 1941. Trading ideas with your neighbors. Pamphlet submitted to the U.S. Farm Security Administration. Carl C. Taylor Papers, Collection 3230, Rare and Manuscript Collections, Cornell University Library, Ithaca NY.

Taylor, Donald C. 1990. On-farm sustainable agriculture research: Lessons from the past, directions for the future. *Journal of Sustainable Agriculture* 1 (2): 43–87.

Thornley, Kay. 1990. Involving farmers in agricultural research: A farmer's perspective. *American Journal of Alternative Agriculture* 5 (4): 174–77.

Thrupp, Lori Ann. 1989. Legitimizing local knowledge: From displacement to empowerment for Third World people. *Agriculture and Human Values* 6 (3): 13–24.

Todd, John H. 1971. Shaping an organic America. *Organic Gardening and Farming* 18 (9): 50–55.

U.S. Bureau of the Census. 1995. *1992 Census of Agriculture*. Washington DC: U.S. Department of Commerce.

Vogeler, Ingolf. 1986. *Wisconsin: A geography*. Boulder: Westview Press.

Voisin, Andre. [1959] 1988. *Grass productivity*. Trans. Catherine T. M. Herriot. Covelo CA: Island Press.

Wainwright, Hilary. 1994. *Arguments for a new left: Answering the free market right*. Oxford: Blackwell.

Warren, Dennis Michael. 1994. Indigenous agricultural knowledge, technology, and social change. Chap. 2 in *Sustainable agriculture in the American Midwest: Lessons from the past, prospects for the future*. Edited by Gregory McIsaac and William R. Edwards. Urbana: University of Illinois Press.

Weick, Karl E. 1984. Small wins: Redefining the scale of social problems. *American Psychologist* 39 (1): 40–49.

Whatmore, Sarah. 1988. From women's roles to gender relations: Developing perspectives in the analysis of farm women. *Sociologia Ruralis* 28 (4): 239–47.

———. 1991. Life cycle or patriarchy? Gender divisions in family farming. *Journal of Rural Studies* 7 (1, 2): 71–76.

———. 1994. Global agro-food complexes and the refashioning of rural Europe. Chap. 3 in *Globalization, institutions, and regional development in Europe*. Edited by Ash Amin and Nigel Thrift. New York: Oxford University Press.

Winner, Langdon. 1986. *The whale and the reactor: A search for limits in an age of high technology*. Chicago: University of Chicago Press.

Wisconsin Department of Agriculture, Trade, and Consumer Protection (WDATCP). 1991a. *Farmer-to-farmer networks*. Madison: State of Wisconsin.

———. 1991b. *Sustainable Agriculture Program project directory, 1987–1991*. Madison: State of Wisconsin.

Wisconsin Rural Development Center. 1995. *The grass is greener: Dairy graziers tell their stories*. Mt. Horeb: Wisconsin Rural Development Center.

Youngberg, Garth, Neill Schaller, and Kathleen Merrigan. 1993. The sustainable agriculture policy agenda in the United States: Politics and prospects. Chap. 12 in *Food for the future: Conditions and contradictions for sustainability*. Edited by Patricia Allen. New York: Wiley.

Index

Only those proper names that figure promi-
nently in the theoretical and historical analyses are included here. Individual farmers'
names are not indexed.

agrarianism, 4, 12, 20, 105, 112
agricultural extension agents. *See* extension
agents
agricultural science/research, 10, 12; choice of
questions studied by, 18, 20, 52, 183; critiques of, 14–15, 22, 38, 116, 119, 144–45;
farmers' opposition to, 13, 86; farmers' role
in, 13, 24–25; generalizability of, 16–17; interests served by, 18–22; lack of diversity in,
21–22; lack of holistic perspective in, 17–18;
reform efforts of, 6, 22–26; successes of,
14–21; and transformation of U.S. agriculture, 13–14; worldview assumptions of, 15–
18. *See also* knowledge; land-grant colleges/
universities; rotational grazing: university research on
Allen, Patricia, 3, 5, 26, 186
Appropriate Technology Transfer for Rural
Areas (ATTRA), 23, 145

Berry, Wendell, 4, 6, 20

Campaign for Sustainable Agriculture, 1
Carson, Rachel, 19
Center for Integrated Agricultural Systems,
24, 87
class discrimination, 169, 171
community supported agriculture (CSA), 46–
47, 152, 169
conventional agriculture, 2, 30, 68, 82, 113,
120–21, 145, 174; critiques of, 3–4, 9, 22–23;
and dairy farming, 50, 52, 77, 102–3, 106–7

ecological concerns, 4–5, 14, 18, 20, 24, 165–
67
empowerment. *See* women: and sense of personal agency
environment, conceptions of, 16–17
extension agents, 12–13, 24–25, 41–42, 75–
76, 86–87, 143–45
Eyerman, Ron, 2, 7, 35, 102, 182

farmer knowledge. *See* knowledge: farmer-
generated
farmer networks, 2, 29–30, 35, 38, 43–49,
67; defined, 40; gendered participation in,
33, 53, 55–56, 71, 100–101; and knowledge
creation, 97–98; and knowledge exchange,
2, 6, 27, 30, 38, 42, 46, 52, 58, 66, 73, 178;
organizational characteristics of, 45–46, 52,
56–57, 69, 123–25, 127–28, 180; relation to
universities of, 42; support function of, 8,
68, 113–21, 123, 159, 162, 169–77, 184–85,
187, 189; in Wisconsin, 40–47, 66, 191–92.
See also Ocooch Grazers Network; Wisconsin Women's Sustainable Agriculture Network
farmer-to-farmer organizations. *See* farmer
networks
Feldman, Shelley, 33, 181
feminism, 5–6, 14–15, 31–32, 34, 123, 170,
181, 185; women farmers' identification
with, 160–65, 176, 188